CAMBRIDGE COUNTY GEOGRAPHIES

General Editor: F. H. H. GUILLEMARD, M.A., M.D.

NORTHAMPTONSHIRE

Cambridge County Geographies

NORTHAMPTONSHIRE

by

M. W. BROWN, M.A.

ASSISTANT MASTER IN OUNDLE SCHOOL

With Maps, Diagrams and Illustrations

Cambridge:

at the University Press

1911

CAMBRIDGE UNIVERSITY PRESS
Cambridge, New York, Melbourne, Madrid, Cape Town,
Singapore, São Paulo, Delhi, Mexico City

Cambridge University Press
The Edinburgh Building, Cambridge CB2 8RU, UK

Published in the United States of America by Cambridge University Press, New York

www.cambridge.org
Information on this title: www.cambridge.org/9781107630987

© Cambridge University Press 1911

First published 1911
First paperback edition 2013

A catalogue record for this publication is available from the British Library

ISBN 978-1-107-63098-7 Paperback

PREFACE

I SHOULD like to acknowledge here the help which I have received from friends. The Headmaster of Oundle School has given me assistance in many ways; my colleagues Mr Ll. R. Jones, Mr E. I. Lewis and Mr J. G. Hornstein have given me a great deal of information and have most kindly looked over the proofs. I am deeply indebted to my brother-in-law Mr G. S. Russell who has taken most of the photographs used for illustrating the book, and has helped me in many other matters. Mr T. J. George, Curator of the Museum at Northampton, has rendered valuable assistance, and it is due to the courtesy of the Museum Committee that it has been possible to give illustrations of antiquities found within the county. Mr Alfred Gotch has given his help in the chapters dealing with architecture, and Mr Christopher A. Markham has advised me on points of historical and antiquarian interest. Mr John Siddons,

Surveyor to the Nene Commissioners, has taken considerable pains to put me in possession of important facts with regard to the waterways of the county. Dr H. R. Mill most courteously gave me his advice as to the best way in which the distribution of rainfall in the county might be shown.

I have consulted many books dealing with the county, and I am particularly indebted to the *Victoria County History*.

M. W. BROWN.

August, 1911.

CONTENTS

ILLUSTRATIONS

MAPS

1. County and Shire.

When we look at a map of England we cannot fail to be struck by the way in which the land has been divided up into the different areas which we call counties or shires. It is not easy at first to see what method has been pursued in the division. There are two well-known ways of dividing up a large area of land. One is by the use of regular lines, such as parallels of latitude and meridians of longitude; a method which, though not employed in England, has been used largely in North America and in Australia. The other is to make the dividing lines follow some natural feature of the land whenever that is possible, and this has been the rule in England in many places, but not at all uniformly. For instance, the division into counties in the north of England clearly shows the influence of the Pennines, and in the south the river Thames has played an important part. There is yet another method, that of growth from a smaller unit. History teaches us that the division which we know as a county or shire grew from a collection of smaller units of area, the "hundreds." These hundreds were probably tracts of land inhabited by a hundred families; when larger divisions were wanted these were obtained by grouping together a certain number of hundreds; and just as the outside boundaries of certain properties would form the boundary of the hundred, so the outside boundaries of certain hundreds would form the boundary of the shire. In fact the county was formed out of a certain number of hundreds grouped together, and was not, as might be supposed, subdivided into hundreds. The boundary of the shire might or might not follow natural features of the country; the determining factor was the boundary of the smaller unit.

In this book we are going to deal with one particular county, Northamptonshire, so that our attention will be riveted on a small tract of land, the size of which is about one fifty-eighth of that of England and Wales.

The first thing to do is to consider the name given to this division of the country. If we look at the names of other counties, we at once notice that the ending -*shire* is common to a great many. This word *shire* is derived, like the word *share*, from an old English word meaning "to shear," "to cut off," so that it is clear that the name Northamptonshire means "the division (or share) of Northampton." We might at once guess that the town of Northampton must have existed before the division was made of the land which gave a certain portion of the country to it as its "shire." We should be right in this, for although we do not know for certain when Northampton was actually founded, we do know that it had been in existence for some centuries before any division such as that of "shires" was effected: we find it mentioned in the Saxon Chronicle as Hamtune, and it must have received later the prefix North- to distinguish it from Southampton, which is also referred to in the same chronicle as Hamtune. Early in the tenth century the town was held by the Danes, and in 1010 Sweyn almost completely destroyed it. Ever since that time it has been a town of importance, and several parliaments have been held within its walls. But we shall have more to say of its history later.

We have learnt, then, the meaning of the word Northamptonshire. It remains for us to find out when the division took place which marked off our county as "the share of Northampton."

If we look at a map of England after the Treaty of Chippenham in 878, we shall not see any district marked Northamptonshire, but we notice that the boundary of Danish Mercia, which divides it from the kingdom of Guthrum and from English Mercia, follows roughly a great part of the boundary of our modern county. In 878 then, we may say, Northamptonshire formed the southern portion of Danish Mercia. In all probability it was at some time during the century succeeding the Treaty of Chippenham that England was divided up into counties or shires. Many of the shires took their names from the chief towns in them, so that it was only natural that this county should receive the name of Northampton-shire.

Following the revolt of the Northumbrians in the reign of Edward the Confessor, a fresh division was made of the kingdoms in the north, and we find that Northamptonshire and Huntingdonshire were assigned as

The Nene near Elton

(*The boundary between Huntingdonshire and Northamptonshire*)

a separate earldom to Waltheof. We also find the county mentioned in Domesday Book, when it included much of Rutland, but in the reign of Henry II it was reduced practically to the area and shape which it has now.

2. General Characteristics. Position and Natural Conditions.

When we turn to examine the general characteristics of our county, the first thing that we notice is that it is what is termed an inland county, that is to say it is bounded on all sides by land. It stands moreover in that part of England which is termed the Midlands, a little to the south-east of the centre. It is difficult to see any natural boundaries ; but if we look at a map which shows the height of the land (such as that at the beginning of this book), we cannot help noticing that most of Northamptonshire seems to form the upper part of the basin of the Nene. The county indeed rather resembles a leaf in shape, with its higher ground in the north-west and south-west, and with the mid-rib of the leaf represented by the river Nene.

There is an important point to be noticed with regard to its position. From the earliest times the great roads—and in these days of our own the railways—have radiated from London in every direction. Now to reach either Scotland or the north of England from London, one must decide to go either to the east or the west of the Pennines, that broad stretch of high ground which reaches down into England as far as Derbyshire. We can, in fact, travel north from London by one of two ways—either through the Midland gate and then through Lancashire, or else to the right of the Pennines by the Vale of York. If we take a ruler and draw straight lines from London to Crewe and from London to York,

View near Arbury Hill
(*Showing hilly ground characteristic of West Northamptonshire*)

we shall find that both these lines cross our county. Northamptonshire, then, lies directly in the way of traffic from London to the north by either of these two great routes. Its position therefore has made it an important thoroughfare on the way from south to north—indeed four main railway lines cross the county.

View across Nene valley near Oundle

(*Showing the undulating lands of East central Northamptonshire*)

The shape of Northamptonshire has given it traffic in another direction, from south-west to north-east. Between London and the Pennines England is crossed by bands of high ground which run, roughly speaking, from south-west to north-east. Communication therefore between the west and east of central England has at all times naturally followed the lower ground in between these bands of high ground, and consequently the main valley of our county, that of the Nene, has been much used as a route from west to east.

It is usual to take a line drawn from the Severn mouth to the Wash as dividing industrial north-west England from the south-east part which has no great industries. Northamptonshire lies just to the south of this line, and is therefore not included in what is termed "industrial England." As a matter of fact it possesses industries, some of them very important ones, but on the whole it is an agricultural county. It has good soil, an equable climate, and a moderate to low rainfall ; and as nearly all the land is capable of cultivation it would be curious if agriculture did not thrive.

The north-east corner is part of the low-lying flat district known as the Fens, but most of the county is undulating, and in parts it is hilly, although it does not rise much above 700 feet anywhere. Most of the highest ground is in the south-west, the west, and along the north-west border of the county, and in these parts the finest scenery is to be found. In many districts there are fine woods, and the uninteresting appearance of the land at the east end of the county is frequently improved by

the presence of noble trees which break the monotony
of the view. In the rainfall map (see p. 51) in which
the contours for 200 ft. and 400 ft. are drawn, it is
possible to see at a glance the three chief districts out-
lined above, the flat land of the east, the high ground
of the west, and the undulating country in the middle
of the county, each having its own strongly marked
characteristics.

3. Size. Shape. Boundaries.

As we have seen, then, in the preceding chapter,
Northamptonshire is shaped somewhat like a leaf, with
its stem pointing to the north-east. This north-east
corner approaches more closely to the sea than any other
part of the county, reaching to within 18 miles of the
wide bay or gulf which we know as the Wash.

The length of the county from south-west to north-
east is between 67 and 68 miles. The breadth of course
varies, being greatest to the south-west of the centre ;
a line drawn across the county, following the direction
of Watling Street for a greater part of the way, is
29 miles in length.

The area of the county (including the Soke of
Peterborough) is 638,612 acres, or nearly 1000 square
miles. If we compare Northamptonshire in size with
Yorkshire the largest county in England, and with
Rutland the smallest, we find that, roughly speaking, it
is one-sixth the size of Yorkshire and six and a half times

the size of Rutland. If we divide the total area of England and Wales by the number of counties into which the land is divided, we find that the average size of a county is about 1116 square miles, i.e. a little greater than the size of Northamptonshire.

The county is bounded on the north by Leicestershire, Rutland, and Lincolnshire; on the east by Cambridgeshire, Huntingdonshire, and Bedfordshire; on the south by Buckinghamshire and Oxfordshire; and on the west by Warwickshire. Thus its borders are touched by nine other counties, a larger number of bordering shires than can be claimed by any other county in England.

If we start in the south-west corner of Northamptonshire and trace the boundary northward, we find that it coincides with the course of the river Cherwell to a point a little to the north of Banbury; here it turns east, and soon after passing Chalcombe bends in a north-west direction, crosses the Cherwell, and after an irregular course on high ground runs down to the Leam, which it follows as far as Braunston. A little to the north of this point it bends to the west, then follows Rains Brook to the east for a short distance, and a little to the north of Kilsby joins Watling Street, which it follows to the point where the Roman road crosses the Avon. Here it turns north-east and follows the Avon almost to its source. Near Welford it crosses the high ground separating the basins of the Avon and Welland, and follows the latter stream almost continuously, past Stamford and Market Deeping to a point about two miles west of Crowland. Here it turns south-east, and after a few

bends crosses the Fen country in a southerly direction
to Peterborough, where it joins the Nene south of the
cathedral city, and follows the course of the river almost
without a break as far as Elton.   At this point it bends
away to the south-east and rises to the higher ground
forming the watershed between the basins of the Nene
and Ouse.   Roughly speaking it follows this watershed
until it passes Yardley Chase.   Two miles south-east
of Stoke Bruerne it reaches the river Tove, when it turns
south, following this river as far as its confluence with the
Ouse.   Here the boundary turns south-west, coinciding
with the course of the Ouse as far as Thornton ; it then
turns north-west, crosses the higher ground which separates
the upper valleys of the Tove and Ouse, and descends
once more to the latter river, the course of which it follows
for a short distance before it bears away to the south and
west towards the valley of the Cherwell near Aynho.
It should be noticed that the seven rivers Nene, Tove,
Ouse, Cherwell, Leam, Avon, and Welland play a large
part in defining the boundaries of the county.

In stating the area of Northamptonshire we included
that of the Soke or Liberty of Peterborough.   These are
the names given to a portion of the county in the north-
east, covering an area of a little more than 83 square miles.
This district was in 1888 declared an administrative county,
separate from the rest of Northamptonshire.   For the
purposes of parliamentary representation, however, it is
taken in with the remainder of the county.   "Soke"
itself is a form of the word "Soc," which meant in old
English times both the privilege of holding a court and

The Nene, as a county boundary, near Elton

the district held by tenure of "socage," a term which denoted the service of freemen who held land in return for which they were bound to render certain services. "During the Saxon period the lord of the soke of Peterborough had the power or liberty of holding a court and administering justice within its boundaries, and this system was subsequently continued by the Abbots of Peterborough, who either enforced in person, as lords, the observance of the ancient socage laws and customs, or appointed a deputy to act for them.   On the establishment of quarter sessions (1349–50), the separate jurisdiction of the soke was still maintained as distinct from that of the county of Northampton ; and except for parliamentary purposes and matters relating to the militia, it is entirely independent of that county " (*Kelly's Directory of Hunts. and Northants.*).   We shall have to refer again to the Soke when treating of the methods of Administration in our county.

4. Surface and General Features.

To gain a thorough knowledge of any division of land it is essential that we should know all the chief characteristics of its surface—its hills and valleys, its tracts of woodland and fen.   We all know the chief points of interest in connection with our own town or village and the neighbouring country ; there are hills, perhaps isolated, perhaps forming part of some upland which stretches away to parts which we have never yet visited ; we may have a river whose course to the distant sea we have traced on a map, or only some small streams flowing away to join some river in the neighbourhood. We learn also to distinguish between the land in one part and the land in another; in one direction it may be bleak and poor, only fit to be used for rough pasture; in another rich and well worth cultivating; in yet another low-lying and marshy. To understand our county properly, we must know its surface in the same way as we know that of the particular district in which we live; we shall then be better able to understand the differences to be observed in the occupations and pursuits of the people who live in different parts of the shire.

The first fact, which it is important that we should grasp, is that the land stretching from London in a north-west direction was at some remote time in the past tilted, so that, roughly speaking, it sloped steadily downwards in a south-east direction. The parts raised highest suffered most from denudation, and that is the reason why, if we travel from London in a north-west direction, we find strata of rock ceasing suddenly, and layers of older rock appearing at the surface; first the layer of clay, on which London stands, ceases; then the chalk appears, to be followed by the greensand, and finally the rocks of the Jurassic group appear, which have weathered into the undulating country characteristic of the county of Northampton. With these strata one peculiarity is to be observed; their slope on the south-east side (their "dip") is gentle, that on the north-west side (their "escarpment") is comparatively abrupt. This can be clearly seen in an orographical map of Northamptonshire. Between the valley of the Nene and that of the Welland you can trace from south-west to north-east a belt of high ground, whose sides slope downwards to the Nene in the south-east and to the Welland in the north-west. Along the highest part of this belt runs the watershed between the two rivers, dividing the basin of the one from that of the other. Now it can be easily seen that this watershed is much nearer to the Welland than to the Nene, and therefore, as the bed of the Welland is little higher than that of the Nene, the slope of the high land dividing the basins of the two rivers is steeper on the north-west side than on the south-east.

For the sake of convenience we may divide the surface of our county into six parts :—(1) a hilly portion in the south-west ; (2) a band of high ground running from Daventry in a north-east direction and ending rather abruptly near Stamford ; (3) a belt of ground, following roughly the south-east border of the county, not very high, but forming the watershed between the Nene and the Ouse ; (4) the north-east corner of the county, comprising the Soke of Peterborough, mostly fen-land ; (5) the valley of the Nene, lying between (2) and (3) ; (6) the southern half of the valley of the Welland, which runs for some distance along the north border of the county. We can now consider some of the most important points to be noted in connection with each of these divisions.

The first and second may be considered together. They include all the highest land of the county ; their surface is rounded and undulating, all the sharper features

View from the top of Arbury Hill showing the head of the Nene Valley

of high lands having been removed by the long process of denudation to which they have been exposed. They are united by a narrow strip of high ground, near Daventry, over 400 ft. in height, and together they form the most important water-parting in central England. Here are to be found the sources of the Avon, flowing west and south-west to join the Severn; of the Cherwell, flowing south to join the Thames; and of the Welland, Nene, and Ouse flowing east and north-east to the Wash. In no place does the land rise to any considerable height; Arbury about 800 ft., Charwelton Hill over 700 ft., and Naseby over 600 ft. are amongst the highest points. It should be noticed that the band of high ground on the north-west side of the county becomes considerably lower as we trace it in a north-east direction; no part, indeed, north of Rockingham rises to a height of 400 ft., and the high ground ceases altogether near Stamford. The whole district forms part of the great band of Jurassic limestone which runs across England from the Cotswolds to the Cleveland Hills in Yorkshire.

The third division, the belt of ground which divides the basin of the Nene from that of the Ouse, and follows the south-east border of the county for some distance, calls for little comment. Much of it is over 200 ft., but in no place does it reach a height of 400 ft. It is only of interest, from a geographical point of view, as forming a watershed between two important river basins, and affording a natural boundary between our county and the shires which lie to the south-east. To the geology of the district we shall have to refer later.

In the north-east corner of Northamptonshire, our
fourth division of the surface, we find a portion of the
district known as the fens. This is a low-lying region,
with a flat surface, the monotony of which is only broken
by occasional clumps of buildings and trees. Now that
it has been drained the district is a fertile one, and
agriculture flourishes.

The Nene in flood

We have already pointed out the only feature in the
valley of the Welland worthy of comment in connection
with our county, namely the comparatively abrupt descent
from the watershed which divides it from the valley
of the Nene. The latter, however, which forms our
fifth division, is one of the most important surface features
of the county. It runs from Arbury Hill in the west to
Wansford, where it ceases to be a distinctive feature of

the landscape, because at that point it opens upon the
low-lying flat land which extends to the Wash.   It forms
the great highway of communication between the two
most important towns of the county, Northampton and
Peterborough.   The valley marks out the route of communication by road and rail between the south-west and

Causeway across the Nene Valley at Oundle

the north-east, and the river itself affords a means of water
traffic—little used now, but of importance in the past.
Broad flat meadows, yielding good pasture, with gently
undulating hills to the north and south which are crowned
in many places with woods and bear upon their sides
or summits venerable churches affording some of the
finest examples of ecclesiastical architecture ; long stone

2—2

causeways traversing the valley at right angles, built to
keep the roads open to traffic in times of flood ; these are
the principal features which strike an observer as he travels
down the valley.   The course of the river Nene itself we
shall consider in the next chapter.

5.   Watershed.   Rivers   and   their Courses.

If we look at a map of England in which the land
is divided up, not into counties, but into the different
catchment   basins   from   which   the   rivers   derive   their
waters, we cannot fail to be struck with the resemblance
in general shape between the basin of the Nene and the
county of Northampton.   As will have been gathered
from the preceding chapters, Northamptonshire is not
identical with the basin of the Nene.   In the first place,
the Nene flows out of the county and touches or passes
through other counties before it reaches the sea.   Northamptonshire therefore does not contain the whole of the
Nene basin.   Again, the high land in the south-west
corner, and running along the north-west edge of the
county, forms the watershed between the basins of the
Nene and other rivers ; the band of high ground also
along the south-east side of the county separates the
basins of the Nene and the Ouse.   Northamptonshire
therefore contains parts of the basins of other rivers
besides the Nene.   The greater part of the county,
however, is occupied by land which is drained to the

Nene, and a great part of the Nene basin is comprised
within the county.

We can examine separately the parts of our county
which fall within the basins of different rivers, and as the
Nene is by far the most important to us, we will turn
our attention to it first.

The Nene rises near Staverton, at a height of a little
over 500 ft., and flows practically due east through
Weedon Beck, where the Grand Junction Canal crosses
it, to Northampton.   Here, on its left bank, it receives
from the hills to the north its first tributary of importance, one of whose branches has its source near Naseby,
while the other, the Stowe Brook, rises near Cold Ashby.
From this point eastward the river is navigable for barges.
After passing Northampton, the Nene bends northward
and flows in a north-west direction to Wellingborough,
where it receives, on its left bank, its most important
tributary, the Ise.   The latter has its source in two
streams, one of which, the northern, rises near Sulby
Hall, to the south-west of Sibbertoft, at a height of over
500 ft.; this stream flows east, past the site of the battle
of Naseby, and at Arthingworth is joined by the second
stream which has risen near Naseby at a height of nearly
600 ft., at a point within two miles of the source of the
tributary which joins the Nene at Northampton.   The
Ise takes a southerly turn at Rushton, and flows past
Kettering to join the Nene to the south-east of Wellingborough.   After passing this town the Nene turns more
to the north; it flows past Thrapston, receiving soon
afterwards from the north the waters of Harper's Brook;

it continues its winding course past Oundle, where it makes a great loop, and, after passing Fotheringhay, is joined on its left bank by the Willow Brook. From Elton the river flows north to Wansford, where it turns to the east and runs past Water Newton to Peterborough. It is worth noticing that from Elton to Peterborough

Weir on the Nene, Elton

the boundary of the county practically follows the course of the Nene.

There are two special points of interest in connection with the main stream and its tributaries. In the first place the Nene receives no tributaries of importance on its right bank. The cause of this is clear when we remember, firstly, how near the river flows to the

watershed in the south-east which separates its basin from
that of the Ouse, and secondly how low that watershed is.
In the second place the courses of three of the four left-
bank tributaries bear a marked resemblance to one another.
The Ise and Harper's Brook both flow east at first, then
south-east to join the Nene ; the Willow Brook only

Mill on the Nene, Barnwell

differs from the other two in that, after starting in an
easterly direction, it turns north-east before bending south-
east to join the main stream.   The similarity of these
stream-courses draws our attention to the fact that from
the belt of high land in the north and north-west of the
county spurs of high ground jut out eastward, and the
valleys between have been trenched out in the past by the

rivers flowing down to the Nene valley. The lower parts
of these valleys are very clearly marked in the rainfall map
(on p. 51) on which the 200 ft. and 400 ft. contours are
shown.

The banks of the Nene are low; consequently, if the
river rises a little, it floods the adjoining land. Moreover
the land on each side is very flat, so that the floods,

The Nene in flood
(*The water has invaded the granaries on the right*)

although shallow, are apt to be large in extent. This
liability to flooding has had two direct consequences;
firstly, the roads crossing the valley are built up to a
sufficient height to make traffic by road across the valley
always possible; and secondly, the villages and towns
which lie in the valley are, as a rule, built at a little

distance from the river on rising ground which no ordinary
flood can reach.   This flooding of the river has a distinctly
beneficial effect upon the land, for the waters, when they
subside (which they do almost as quickly as they rise),
leave, spread over the land, the silt which they held in
suspension.   It is to be noticed that the floods generally
reach their highest level one or two days after there has

The Nene in flood

been a heavy rainfall in the west of the county, that is
to say the river, in its lower parts, only feels the full effect
of the rain which has fallen in its upper reaches some
time after the rainfall has ceased.

We must point out one more matter of interest with
regard to our chief river ; its source lies at a height of

500 ft. above sea-level, and the length of its course to the sea is about 100 miles. Of the fall of 500 ft. from source to mouth, the river drops 300 in its first 14 miles, and for the last 50 miles (i.e. during the second half of its course) it falls only 50 ft. Hence the slowness of the river and the ease with which it can be navigated by barges.

We can now turn to those portions of our county which form parts of the basins of other rivers. On the north-west border of Northamptonshire the Welland rises, near Sibbertoft, and flows west for a short distance until it reaches the county boundary, when it turns, rather abruptly, to the north-east and flows along the north-west boundary; here and there it leaves the boundary, but it follows it fairly closely for many miles until, some 12 to 14 miles after passing Stamford, it finally turns north-east and north when it is nearing Crowland, and bears away to the sea. Like the Nene, and for a similar reason, the Welland has no tributaries on its right bank.

The next river system connected with our county is that of the Avon. This river, like two of the Nene's tributaries, rises near Naseby, but on the north-west side of the hill; it flows at first in a north-west direction, then south-west along the border of the county, which it leaves at the point where Watling Street crosses it. It receives later the waters of the Leam, which rises near Hellidon, just within the borders of our county.

The south-west corner of Northamptonshire falls within the basin of the Thames, for the Cherwell, after rising near Charwelton within the county, flows first south, then west, then south again past Banbury, its course coinciding with the boundary of the county in its extreme south-west corner.

We pointed out in Chapter 3 that the south-east edge of Northamptonshire practically follows the watershed between the Nene and the Ouse. The southernmost portion of the county, however—roughly speaking the district round Towcester and to the south of it—actually forms part of the basin of the Ouse. The river Ouse itself rises within Northamptonshire, north-north-west of Brackley, although it leaves the county almost immediately. It is indebted however to Northamptonshire for the waters of the Tove which, rising near Sulgrave and flowing through Towcester, joins the larger stream $1\frac{1}{2}$ miles north-east of Stony Stratford.

It should now be perfectly clear why we stated in the preceding chapter that the high ground in the south-west, west, and north-west of the county forms the most important watershed in central England. Within the county borders lie portions of the basins of the Nene, the Welland, the Severn, the Thames, and the Ouse. It is interesting to notice how much of the boundary of Northamptonshire is determined by rivers. Every one of the rivers whose courses we have traced above forms a part of the boundary of the county at some point or other of its course.

6. Geology and Soil.

Soil is formed of two things, broken-up rock and decaying organic matter. By the word "rock" geologists mean all substances which compose the crust of the earth. Before we consider the geology of Northamptonshire, we must try to understand something about the general characteristics of rocks.

At some time in the past, many geologists believe, all the rocks forming this planet must have been so hot that they were in a molten state. As the earth cooled, an outer skin of solid rock was formed, and this outer covering must have grown thicker and thicker in proportion as the cooling of the earth went on. Solid rocks formed in this way are called "Igneous." As the cooling continued, disturbances took place which made the solid crust crumple, thus giving rise to tracts of greater and less altitude. Directly this first type of rock had been formed, exposure to the atmosphere rendered it subject to the same kind of wear and tear as that to which the rocks of the earth's surface are subject to-day. They were worn away, and the particles broken off collected in the hollows, and were swept into the seas by the rivers ; here they tended to become compact masses once more, owing to the pressure exerted on them by the masses accumulating above, and eventually formed a second type of rock, which we term "Sedimentary" or "Stratified," because it is laid down in *strata* or layers. There is a third great class of rocks, to which the name "Metamorphic" is

	Names of Systems	Subdivisions	Characters of Rock
TERTIARY	Recent Pleistocene	Metal Age Deposits Neolithic ,, Palaeolithic ,, Glacial ,,	Superficial Deposits
	Pliocene	Cromer Series Weybourne Crag Chillesford and Norwich Crags Red and Walton Crags Coralline Crag	Sands chiefly
	Miocene	Absent from Britain	
	Eocene	Fluviomarine Beds of Hampshire Bagshot Beds London Clay Oldhaven Beds, Woolwich and Reading Thanet Sands [Groups	Clays and Sands chiefly
SECONDARY	Cretaceous	Chalk Upper Greensand and Gault Lower Greensand Weald Clay Hastings Sands	Chalk at top Sandstones, Mud and Clays below
	Jurassic	Purbeck Beds Portland Beds Kimmeridge Clay Corallian Beds Oxford Clay and Kellaways Rock Cornbrash Forest Marble Great Oolite with Stonesfield Slate Inferior Oolite Lias—Upper, Middle, and Lower	Shales, Sandstones and Oolitic Limestones
	Triassic	Rhaetic Keuper Marls Keuper Sandstone Upper Bunter Sandstone Bunter Pebble Beds Lower Bunter Sandstone	Red Sandstones and Marls, Gypsum and Salt
PRIMARY	Permian	Magnesian Limestone and Sandstone Marl Slate Lower Permian Sandstone	Red Sandstones and Magnesian Limestone
	Carboniferous	Coal Measures Millstone Grit Mountain Limestone Basal Carboniferous Rocks	Sandstones, Shales and Coals at top Sandstones in middle Limestone and Shales below
	Devonian	Upper } Devonian and Old Red Sand- Mid } stone Lower }	Red Sandstones, Shales, Slates and Lime- stones
	Silurian	Ludlow Beds Wenlock Beds Llandovery Beds	Sandstones, Shales and Thin Limestones
	Ordovician	Caradoc Beds Llandeilo Beds Arenig Beds	Shales, Slates, Sandstones and Thin Limestones
	Cambrian	Tremadoc Slates Lingula Flags Menevian Beds Harlech Grits and Llanberis Slates	Slates and Sandstones
	Pre-Cambrian	No definite classification yet made	Sandstones, Slates and Volcanic Rocks

given ; this class consists of rocks which have been sub-
jected to such heating, or pressure, or chemical action
(not necessarily all at the same time), that their nature
has been changed ; it is indeed often difficult to determine
to what group of rocks they should be assigned.

The surface rocks of Northamptonshire are all of the
type known as sedimentary. We have already referred
to the tilt which the land must have received ages ago,
and which has considerably influenced the configuration
of this county. This tilt has revealed to us the rocks,
of which the crust of the Midlands is composed, in the
order in which they lie one upon another, for as we move
in a north-west direction we come to one new rock after
another, the masses that once lay above them having been
removed by long ages of denudation. From the south-
west of Northamptonshire to the north-east can be traced
bands of rock, which as a rule rise gently from south-east
to north-west and then cease, rather abruptly, leaving
some new rock exposed. We must consider these
bands of rock in order and try to get some idea of the
characteristics of each, turning to the geological map of
the county at the end of this book, and to the section of
the county on the opposite page.

Firstly, from Rushden to a little beyond Warmington
there stretches a broad band of Oxford Clay, capping the
hills which form the watershed between the basins of the
Nene and the Ouse. Facing this, but on the opposite
side of the Nene, there is a large tract of country
stretching from Aldwinkle to Cotterstock Wood covered
with the same rock ; this originally formed one continuous

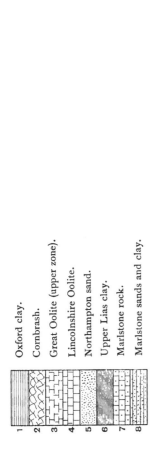

1 Oxford clay.
2 Cornbrash.
3 Great Oolite (upper zone).
4 Lincolnshire Oolite.
5 Northampton sand.
6 Upper Lias clay.
7 Marlstone rock.
8 Marlstone sands and clay.

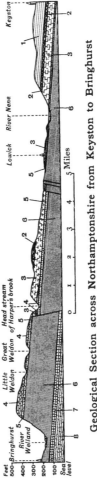

Geological Section across Northamptonshire from Keyston to Bringhurst

(Notice vertical exaggeration ; the dip is very slight, the beds being almost horizontal. The Boulder Clay, which is found in most parts, is not shown above)

layer with that to the south-east of the river, but was cut off by the river Nene gnawing its way through. Isolated patches of this clay are found in a few other places, e.g. in Brigstock Park and in parts of the Soke of Peterborough. Such isolated portions of a rock, cut off from the main layer of which they originally formed a part, are called " outliers."

Typical Northamptonshire Cornbrash Boundary Wall

We come next to the uppermost rock of the Great Oolite series, a strip of Cornbrash. This lies directly beneath the Oxford Clay, and therefore fringes it whenever the latter ends. Accordingly we find it exposed along the edge of the Oxford Clay, both on the south-east side of the county and round the outliers. In some cases it forms an outlier by itself, because the whole of the Oxford Clay, which originally rested on it, has been removed by denudation. Cornbrash is a very hard limestone ; when it has been exposed to weathering it breaks up into flat pieces of a brownish tint, and it is consequently largely used for making rough walls. On account of its hardness it has also been used to a great extent for roadmaking.

After the Cornbrash we come to the broad band of Great Oolite proper, which extends almost without a break from the south-west corner of the county to the north-east, peeping out from beneath the Cornbrash, sometimes in a narrow band, but more often covering great areas. The name "Oolite" describes the nature of the prevailing rock, that is to say a limestone consisting of tiny rounded grains, resembling those of a herring-roe, compacted together with a cement of the same substance. This rock is often used as a building stone, being very easy to work. Clay occurs in the same formation.

The next rocks to appear are the Upper Estuarine series of clays. These form a very narrow band of land to the south-east of the Nene, but cover rather larger tracts to the north-west of the river, especially in the centre of the county. These clays are used for brickmaking, but are a hindrance to agriculture.

After this series we meet with the Lincolnshire Oolite, which occupies a large area in the northern half of the county. Many famous stone-quarries are to be found in this district. In places the lower beds yield the so-called slates of Colly Weston, which are used so extensively

for roofing purposes. These, however, are not true slates
—that is to say compacted muds such as are quarried in
Wales—but laminated oolite limestone.

Ironstone Quarry at Islip

a   Soil and subsoil from boulder clay.   *b*   Sands with clay.
c   White sand.   *d*   Blue and brown stiff clay.   *e*   Ironstone
(ferrous carbonate weathering to ferric oxide)   *f*   Clay and
sand removed from above the ironstone and brought over on the
boards shown.

*It is characteristic of this district for the nature of the overlying
rock to change entirely within short distances, by thinning out
and by faulting.*

The next series of rocks is known as the Northampton
Sand, comprising the Lower Estuarine beds, the Variable
beds, and, at the base of the series, the Ironstone beds.

This series changes rapidly in character within short distances. Perhaps the most interesting feature is the occurrence of Ironstone beds of the valuable ore called " Brown Hematite," known to chemists as hydrated ferric oxide, which is very like ordinary iron rust. In places, however, its usefulness is impaired by the presence of too great a proportion of phosphates ; this renders the iron extracted unsuitable for conversion into steel. These last two formations—the Lincolnshire Oolite and the Northampton Sand—are classified together as the Inferior Oolite.

Below the Oolitic Series we meet the Liassic formations, of which the uppermost is the Upper Lias Clay. This extends along the north-west border of Northamptonshire. In the centre of the county, round Kettering and more extensively in the south, it is exposed in fantastic outlines by the lowering of the beds of the tributaries of the Nene. The rock consists chiefly of clays with some limestones.

The middle of the three Liassic formations is the Marlstone which is found in the south-west of the county. This consists of clays and limestones, and mixtures of the two which are called Marls. The soils formed by this rock are highly productive.

Beneath the Marlstone, and exposed to the north and west of it, is the Lower Lias, a formation of greater importance in the counties to the west of Northamptonshire. It extends along the west border of our county, and down the uppermost part of the valley of the Nene as far as a fault near Kislingbury.

In the valleys of the Nene and its tributaries, and of the Welland, we find deposits of alluvium ; and at spots south of Northampton, south of Earl's Barton, south-east of Wellingborough, up the valley of the Ise round Finedon station, and in several places near Thrapston, beds of gravel are situated ; these must have been brought down by the rivers in years long gone by from the high lands in which they rose.   In these gravels many interesting remains of vertebrate animals are continually being found.

Finally we may add that over a large part of the county a thin covering of boulder clay is found, i.e. a clay deposited by an ice-sheet which must, at some time in the past, have extended over a great part of Northampton-shire.   The survey of this feature, however, is not yet complete.

7. Natural History.

Geology shows us that the surface of the earth has at various times undergone alterations and changes of a stupendous nature.   Great mountain chains have been elevated, new seas have been formed, vast land-masses have disappeared.   Some of these changes are exemplified in our own country.   We know, for example, that at one time—and that, geologically speaking, at no very remote period—Britain was not insular, but formed part of the European Continent.   In the caves of Derbyshire and Devon and many other places are found the bones of extinct animals identical with those disinterred in neighbouring continental countries, or dredged up from the bed of the North Sea. Again, in many places around our coasts are to be seen, at extreme low water, the remains of forests buried beneath the sea. Though just off the west coast of Ireland the sea bottom sinks rapidly to very deep soundings, the North Sea is everywhere very shallow, and if London could be placed in it, the dome of St Paul's would be seen standing well above its surface. Great Britain and Ireland are thus examples of what are known to geologists as "recent continental islands," and subsidence, erosion, or other geological changes have turned dry land into the North Sea, and a broad river valley into the English Channel.

But, at some earlier period, when still forming part of the continent, our land was for a time submerged. The existing fauna and flora destroyed by this subsidence had thus to be replaced from the continental lands from the south-east. Slowly these new immigrants worked their way north-westward as the land afforded suitable conditions, but as it was not long before separation occurred and Britain became insular, not all the species existing in the continent were able to establish themselves in our land. We should thus expect to find that those parts of the country nearest the continent were richer in species and those furthest off poorer, and this proves to be the case both with plants and animals. Britain has fewer species than France and Belgium ; and Ireland, which was probably separated still earlier, has fewer than Britain.

Small though our land may be, it is sufficiently large

to exhibit considerable differences in the distribution of
its fauna and flora. These differences are dependent
upon a number of conditions, which are themselves often
mutually interdependent—upon elevation, temperature,
soil, rainfall, and so forth. As in plants, so in animals,
there are some species which are much more widely
distributed and generalised than others, but for the most
part each finds its own characteristic environment, and
the naturalist knows instinctively where he is likely to
meet with it. In Northamptonshire we find no specially
outstanding conditions. The central position of the
county is in a measure against them. Maritime counties are necessarily richer in species than those inland,
and we could hardly hope to find here the varied flora of
Norfolk, for instance, or as many birds as that county
can show, lying as it does in the main stream of migra-
tion. Nor—still speaking of birds—is there the same
chance of stray immigrants from the continent as in the
case of Kent or Sussex, or from America, as in the case
of Cornwall. The mere situation of Northamptonshire
is thus against a large list of plants and animals. Its
physical conditions accentuate this tendency. In spite of
its size it shows no great range in the type of country.
We find an immense preponderance of highly cultivated
land, no lakes, no heathland, nowhere any outcrop of
hard rock, no peat bogs, and no land of any great elevation. One feature is noticeable—the large amount of
woodland, some 25,000 acres in area, although but a
tithe of the vast forests of former days, of Rockingham,
Whittlebury, Salcey, Yardley Chase, and Bedford Purlieus,

and even in these the native woodland plants have now to a great extent disappeared.

There are no large wild animals in our land nowadays, but that there were so formerly is shown by the remains that have been found from time to time in the fens in various parts of the eastern counties. The precursor of our domestic cattle, the great wild ox or aurochs (*Bos primigenius*), roamed in the waste lands and was hunted by man, surviving in a modified form into more or less modern times. At Holdenby or Holmby House was, until the seventeenth century, a herd of the wild white cattle doubtless descended from this stock. The brown bear and wolf were contemporary, and the latter probably existed as late as the fifteenth century. The red deer was wild till the time of Henry VIII or even later, but now both it and the fallow deer are only found in parks. The otter is common, especially in the Nene, and the badger by no means rare, while in a county so celebrated for hunting it is hardly necessary to say that foxes are abundant. Of late the brown Norwegian rat seems to have increased in numbers and has become a great pest. The indigenous or black rat, which it has to all intents and purposes supplanted, is however still to be seen occasionally.

That Northamptonshire can claim the long and varied list of birds that it does is not a little due to the fact that it was the home of one of the greatest of English ornithologists, Lord Lilford, whose enthusiasm in his subject aroused interest in others and enlisted numerous observers. Records of observations have thus been kept over a long period of years, and show some interesting facts. As might be imagined, the low-lying valley of the Nene, with its tendency to flood, has doubtless accounted for the visits of a number of the rarer ducks in addition to the numerous commoner species. The scaup, tufted duck, golden-eye, gadwall, shoveller, and sheldrake all

The Aviaries at Lilford Hall

occur, and even that true salt-water duck, the scoter, has more than once been recorded. Many sea-haunting birds, indeed, probably because of the proximity of the Wash, are from time to time noted, not only various species of gull, birds which fly a long distance inland in search of food, but birds such as the cormorant, shag, Manx shearwater, oystercatcher, puffin, little auk, and

Flamingos in the Lilford Ponds

others. Some of the rarer marsh birds are not wanting. The ruff no longer breeds, as it did not so long ago in the neighbouring Cambridgeshire fens, but it is still occasionally met with, while a fine bittern was shot near Oundle in 1885, and the rare spotted redshank has been more than once obtained. Of the larger birds of prey the peregrine is comparatively common, arriving in the

Ponds at Lilford, with Cranes, Pelicans, and other Waterfowl

autumn ; the buzzard, though formerly common enough, is now of exceedingly rare occurrence, and the osprey no doubt only an occasional visitant. But the rough-legged buzzard has several times been noted, and the beautiful little falcons, the merlin and hobby, are frequently to be seen, the latter nesting regularly.

During an " invasion year " of Pallas's sand-grouse

(1888) some of these birds were recorded in the Lilford
neighbourhood, and of other rare foreign visitors the
waxwing, rose-coloured pastor, and hoopoe have been
obtained, the latter on several occasions. The crossbill,
in spite of the abundant woodland, is not a common bird;
the hawfinch on the other hand, once a rarity, has of late
become much more abundant. The beautiful grey wagtail is also increasing, and Ray's wagtail, which takes its
place in the summer, is a familiar bird. The little owl
(*Athene noctua*), a foreigner introduced by Lord Lilford,
has increased very much and has been shot as far to the
south-east as Cambridge. There are two heronries in
the county, a diminishing one at Althorp Park, and one
at Milton Park, which in 1910 had more than 140 nests.

The list of reptiles and batrachians is a very small
one, comprising only the common snake and viper, the
lizard, and the slow-worm, the latter being rare ; while
the common frog, toad, and newt and great crested newt
are the only batrachians. The rivers chiefly produce
coarse fish, but though trout are plentiful in some of the
streams there are few in the Nene. The smelt goes up
the river as high as Peterborough, and the flounder up to
Lilford. The lampern is common in the Nene tributaries, and the crayfish seems to occur in several places in
the county.

To the lepidopterist Northamptonshire is an interesting county ; it is, indeed, conspicuously rich in butterflies
and moths. In the northern part Castor Hanglings and
Bedford Purlieus have for generations been the favourite
haunts of collectors, and many rarities have unfortunately

been still further reduced or even rendered extinct. Barnwell Wold and Ashton Wold, near Oundle, are scarcely less celebrated localities. Of butterflies 53 species have occurred, and there are some, such as the black hair-streak (*Thecla pruni*) and the spotted skipper (*Hesperia paniscus*), which are peculiarly local and rare. The large blue and Mazarine blue (*Lycaena arion* and *L. acis*), as well as the black-veined white (*Pieris crataegi*), have all become extinct in the county, but the splendid purple emperor (*Apatura iris*) is still common.

The botany of the county is too vast a subject to be entered upon here. For a detailed account of it the reader must turn to the admirable paper communicated by Mr G. C. Druce to the *Victoria County History*. But it will be gathered that, from the reasons already given in this chapter, and from various other causes, the flora of Northamptonshire is not a very large one. Mr Druce states that of the 1350 plants which might be expected to occur in the county, under 900 can be regarded as existing in it. Such widely distributed species as the marsh violet (*Viola palustris*) and the two sundews are not found, and—to take well-known plants only—the rushes *Scirpus caespitosus* and *S. maritimus*, the water-avens (*Geum rivale*), the fen-loving *Myrica gale* or sweet gale, the sedges *Carex laevigata, filiformis*, and *canescens*, are all absent, while *Carex echina* and *rostrata* are very rare. There are but few heaths and heath-loving plants. The high cultivation so widely obtaining has undoubtedly played a considerable part in the diminution of our flora.

8. Climate.

"The Air of this County is exceeding Pleafant and Wholefome, the Sea being fo remote, that it is not infected with its noifome Fumes,...." These are the words of a writer on Northamptonshire in the eighteenth century. They are interesting as showing that the healthiness of our climate has long been recognised, but the reason to which it is attributed is hardly one which we should uphold to-day.

Almost every day of our lives we hear some reference to the weather. By this word " weather " we mean the state of the atmosphere with regard to temperature, wind, moisture, and sunshine. The climate of a place means the average weather that is experienced there ; hence the climate of a place depends in the main on these four conditions—the amount of sun-heat received, the prevalent winds, the quantity of rain which falls, and the amount of sunshine.

It is usual to class climates under two heads, marine and continental. The former, as its name implies, is experienced in places which can be influenced by winds from the sea. In the summer the sea is cooler than the neighbouring land, and, as a result, the winds which blow from it moderate the heat of the land ; in the winter the sea does not become as cold as the land which it washes, and consequently the winds from the sea bring warmth to the land. It is clear then that the effect of the sea is to moderate both the heat of summer and the cold of winter. In a continental climate, on the other hand, this moderating influence of the sea is wanting, and the consequence is that great heat may be experienced in summer and great cold in winter. Now the climate of our islands is a marine one; we are surrounded on all sides by the sea, and the result is that we have neither extreme heat in summer nor extreme cold in winter. But we must bear in mind that the further inland we go the more will our climate differ from a marine climate, and the more nearly will it approach to a continental one. Thus, in the case of Northamptonshire, we cannot expect to have as equable a climate as that which is experienced in Cornwall; on the other hand we are not far enough from the sea to experience great extremes of heat and cold.

We can now turn our attention to the other conditions which affect the climate of Northamptonshire, and examine them one by one. The first of these is the amount of sun-heat that we receive; this depends on several things, the most important of which are latitude, height above sea-level, and the slope of the land. We need not discuss the first two at any length; England lies between 50° and 56° N. latitude, and the difference between the latitude of this county and that of any other is not sufficient to produce any very marked variation between the heat received in one and that received in the other. Again, no parts of Northamptonshire are high enough to experience a climate appreciably colder than in other parts on account of the difference in altitude. With regard to the third, it is well known that land in the

northern hemisphere which has a southerly slope receives
more sun-heat than one with a northerly slope, because
the sun's rays fall more directly on the former than on the
latter. It has already been pointed out that as we cross our
county from south-east to north-west the land, on the
whole, rises slowly and then drops comparatively quickly,

Avenue at Southwick

(*Showing direction of prevalent winds*)

so that there is a general south-easterly slope of ground.
This undoubtedly has a good effect on the climate of the
county.

The next condition which we have to examine is the
nature of the prevalent winds.   The winds which prevail
in any country have a very important bearing on the

climate. It is due to their agency that rain is brought to the land; they may also have a marked effect upon the temperature of the district over which they blow. In England generally, it is stated that the wind blows from the west or south-west two days out of three. That this is no overstatement of the case, as far as this county is concerned, is well shown by the illustration on p. 47, of an avenue of trees in Southwick. The hill which appears in the photograph runs towards Oundle in a south-south-easterly direction, so that the trees, bending to the left, i.e. to the east, show clearly the direction of the prevalent winds. These westerly winds from the Atlantic are mild, and it is largely due to them that our winters are not hard. Winds from the east are not frequent, although we experience them to some extent in the spring. From the north-west and north we receive winds with which we associate driving rain and cold bleak weather.

We can now turn to the question of rainfall. When the temperature of any given portion of the atmosphere drops below a certain point, the air is no longer able to hold the water-vapour it contains, and condensation of this water-vapour takes place. This is the fundamental cause of all rain. There are two main causes for this cooling of the atmosphere; the first is the configuration of the land, the second the ascending eddies of the air which we call cyclones. The rain due to the second of these causes may be general all over the country, or it may fall only in particular districts; these cyclones are due to differences of atmospheric pressure, the causes of which must be sought outside our islands, and therefore present no special point of interest with regard to our county. With the rain, however, due to land elevation we have more concern. When a vapour-laden wind reaches high ground it is pushed upwards; cooling takes place, and some of the water-vapour is condensed and falls as rain. A region therefore which contains hills will receive more rain than a neighbouring one which is low, and the edge of a tableland will receive more than the high flat land behind it. We should expect, then, the west of Northamptonshire to receive more rain than the east, and an examination of the rainfall map on p. 51 will show that this is so. All the land in the county above 400 feet is shaded, that between 400 feet and 200 feet is dotted, and land below 200 feet is left white. It will be noticed that at such places as Naseby and Daventry, situated in the west on high ground, the rainfall is over 27·5 inches; at Weedon Beck, which lies lower, the rainfall is between 27·5 inches and 25 inches; in most of the eastern half of the county the rainfall is between 25 inches and 22·5 inches; while in the north-east corner there is a small district where the rainfall is less than 22·5 inches.

The isohyetals (lines of equal rainfall) shown in the map on p. 51 are based on observations taken for the 25 years 1868–1892. They are reproduced from a map constructed by Dr H. R. Mill, who contributed the article on Rainfall to the *Memoir on the Water Supply of Bedfordshire and Northamptonshire*, published in 1909 by the Geological Survey. Full particulars are given there

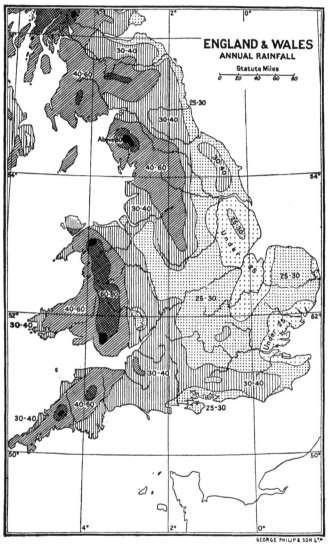

(The figures give the approximate annual rainfall in inches.)

Land above 400 ft. above sea-level.
Land between 400 ft. and 200 ft. above sea-level.
Land below 200 ft. above sea-level.

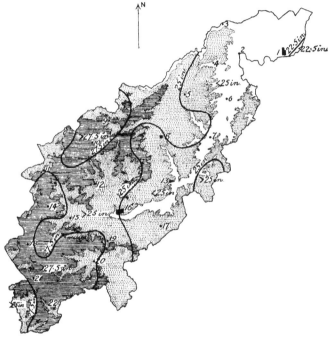

1. Peterborough.	8. Kettering.	15. Weedon Beck.
2. Wansford.	9. Great Oxenden.	16. Northampton.
3. Stamford.	10. Naseby.	17. Yardley Hastings.
4. King's Cliffe.	11. Kilsby.	18. Byfield.
5. Great Weldon.	12. Holdenby.	19. Blisworth.
6. Oundle.	13. Wellingborough.	20. Towcester.
7. Thrapston.	14. Daventry.	21. Thorpe Mandeville.
		22. Brackley.

Rainfall Map of Northamptonshire

Showing regions where rainfall is (1) < 22·5 in., (2) between 22·5 and
25 in., (3) between 25 in. and 27·5 in., (4) > 27·5 in.

The isohyetals traced above are copied, by permission of the Board of Agriculture and Fisheries, from a rainfall map of Northamptonshire prepared by Dr H. R. Mill.

4—2

of the rainfall of our county. It is pointed out that the wettest month is October and that the month with the next highest rainfall is July. March is the driest month.

The last of the four conditions which we have set ourselves to examine is the amount of sunshine that we receive. The two most important conditions affecting this are as follows :—(1) the duration of sunshine is greater near the sea-coast than inland ; the reason is that the land near the sea is, as a rule, low, and clouds form further inland where the ground rises higher ; (2) the duration of sunshine decreases as we go north. We must expect then to have fewer hours of sunshine in North-amptonshire than at places on the sea-coast, and fewer than at places farther south, but we may hope to receive more than many districts lying to the north of us. We find that this is so when we examine a map of our islands constructed to show the duration of sunshine in different parts. The whole of Northamptonshire receives between 1300 and 1400 hours of sunshine in a year, whereas the whole of Cornwall receives over 1500 and most of it more than 1600 hours in a year ; Anglesey receives between 1400 and 1500, and most of Norfolk over 1500 hours a year. Farther north, in practically the whole of Yorkshire, in the whole of Durham and Northumberland, and in most of Scotland, the total duration of sunshine is less than 1300 hours a year.

We may now sum up the chief characteristics of the climate of Northamptonshire ; it is equable, i.e. not subject to extremes of heat and cold, but it is not so voured in this respect as counties bordering the sea, especially in the south and west ; the prevailing winds are mild and blow from the west; the rainfall is moderate in the west of the county where the ground is high, and low in the east where the land is low—this region in fact falls within the area of minimum rainfall for England. The slope of the county is generally favourable to making the most of the sun's rays, and the amount of sunshine that we receive compares favourably with that received by at least half of the kingdom.

9. People—Races, Dialect, Population.

Whatever we know of our county and of the men who lived in it before the invasion of the Romans we owe to the work of archaeologists and geologists. They have marked out certain periods, each of which has received a name from its chief characteristic. The earliest of all is the Palaeolithic or Old Stone Age, when, in all probability, Great Britain still formed part of the continent. Roughly-chipped stone implements, found in gravels within our county, have established the fact that man was in existence here at that early age. We know very little about the physical characteristics of these Palaeolithic men ; on some of their bone implements, however, they have left us sketches of animals.

The next period is called the Neolithic or New Stone Age. The sea meanwhile had carved out the English Channel and the North Sea, and Great Britain had become an island. A fresh race of men, short of stature, long-headed, and with dark hair, had made their appearance, and the stone implements of which they made use showed a decided advance upon those of the preceding age. Such instruments as axe-heads were not merely chipped, but were rubbed smooth until a cutting edge was obtained, and were moreover polished. These men were herdsmen, and they brought with them into this country such animals as the ox, sheep, hog, goat, and dog. Northamptonshire has not yielded many remains of this age. It can have held but few inhabitants; the uplands were covered with dense forests, and the valleys of the Nene and Welland formed large expanses of marsh through which the rivers slowly made their way.

The next period is known as the Bronze Age, from the fact that the new race, who entered the land and overcame the people of the preceding period, introduced the use of bronze. These men were of Celtic origin, rather taller than the Neolithic men, and with rounder skulls; they buried their dead in round barrows and their civilisation was certainly of a higher order than that of the age before them. Our county, judging from the few specimens belonging to this era so far discovered, was still only thinly peopled, and this we may attribute as before to the great tracts of forest and marsh which still covered it.

The last period before the Roman era, where history proper begins, is called the Iron Age. In this period iron superseded bronze as the material used for implements and weapons. Many articles belonging to this age, used by men of Celtic origin like those of the

Bronze Age, have been unearthed at Hunsbury Camp, about two miles south-west of Northampton, and they show conclusively that the men of that day had made a great advance in their use of metal. The first use of coins is also attributed to this age.

The Romans came next, and stayed for about four and a half centuries; they left, however, practically but little mark upon the races inhabiting our island. In the succeeding years, between the departure of the Romans and the invasion of the Normans, the Saxons, followed by the Angles, descended upon Britain. At first they came as pirates, but eventually they settled in the land, the former in the south, the latter in the north, the east, and the Midlands. Both appeared in our county, and their tall, fair-haired, blue-eyed descendants can still be seen. They were destined to form the predominant element in the new race which was to hold Britain. Near the end of the eighth century the Danes appeared, at first as pirates like the Saxons and Angles, but later making permanent settlements. By the treaty of Wedmore, in 878, the Danelagh (or tract in which Danish law was recognised) stretched from the east coast half-way across England, being bounded on the north by the Tees and on the south by the Thames, and embracing the whole of what is now Northamptonshire. These invaders also were a fair race, and eventually they combined with the Saxons and Angles. Their settlements can be distinguished by the termination -by; and of these we have many in the county, e.g. Corby, Ashby, Naseby.

In the middle of the eleventh century the Norman

invasion brought many Normans into our land. These, however, cannot have had much influence on the racial characteristics of the Midlands, which were the result of the fusion of the Saxons, Angles, and Danes with the original British people. With regard to this fusion we must bear in mind the following points; the Angles, who came in greater numbers than the Saxons (and eventually gave their name to our land), entered our county from the north-east and pressed up the valleys of the Nene and its tributaries. The Saxons held the land to the south, and consequently the British would be pushed back westward into the hills, at that time densely forested. Here they would be better able to hold their own, and in time they must have blended with their conquerors into one race. Hence it is not surprising to find that Dr Beddoe has noticed a high percentage of a dark-type race in a group of villages between Weedon and Northampton. Since Norman times there has been no influx of foreigners into our county in sufficient numbers to affect the race in any way.

Northamptonshire lay between Mercia to the north and Wessex to the south. Consequently, to this day, one of the results of this position can be seen in the varying dialects of the county. In the south-west the dialect tends to be like that prevalent in the southern and western counties of England, sometimes known as the west-country dialect; for instance we find *s* pronounced like *z*; the long *a* in *face* pronounced as *ee-a*, turning the word into *fee-ace*; the *ea* in *break* is split into two distinct sounds, so that *break* becomes a word of two syllables *bre-ak*; the *oi*, in such words as *spoil*, becomes *wi*, so that *spoil* is pronounced *spwile*; *ee* often becomes *i*, so that *feet* is pronounced *fit*; *oo* becomes *u*, *took* being pronounced *tuk*; and a common change is that the long broad *o* in *morning* becomes an *a*, giving us *marnen*. It must not be thought that these characteristics are as decided as in the south-west counties; they exist, however, in the south-west corner of our county, although in a less marked degree, and they serve to show the influence of position.

In the north and the east of Northamptonshire the dialect draws closer to that of East Anglia. There are local differences of course; the districts near Leicester-shire have peculiarities connecting their dialect with that of the neighbouring county, and it is the same with the districts which lie near Lincolnshire and North Bedford-shire. The chief points worthy of notice are the pro-nunciation of *i* as *oi*, making *mile* into *moile*; also *a*, when it has the sound of *o* as in *wash*, becomes *ai*, giving us *waish*.

The old plural ending in *-en*, is still often heard; it is not uncommon, for instance, to hear a reference to *housen*. The present writer has also continually heard a curious plural for words ending in *-st*, e.g. *frosses* for *frosts*, and *wrisses* for *wrists*. The glossaries of Sternberg and Baker give full lists of the peculiar words and idioms to be found within the county, and are well worth consulting.

In the middle of Northamptonshire the Anglian speech blended with the Saxon, and the consequence is that the English spoken there has long been famous for its purity. There is a modern tendency, of course, to obliterate slowly the differences of dialect observable in different parts of the country. Cheap and rapid modes of transit have rendered the movement of population so easy that a general "levelling-up" of pronunciation and customs has been discernible for some time. The claim made for purity of pronunciation in Mid-Northampton-shire, however, is an old one. Fuller, when he described the county at the beginning of his work on *Northampton-shire Worthies*, said :—

"The language of the common people is generally the best of any shire in England. A proof whereof, when a boy, I received from a hand-labouring man herein, which since hath convinced my judgment. 'We speak, I believe (said he), as good English as any shire in England, because, though in the singing Psalms some words are used, to make the meeter, unknown to us, yet the last translation of the Bible, which, no doubt, was done by those learned men in the best English, agreeth perfectly with the common speech of our country.'"

Thus we see that the claim that the race found in Northamptonshire is predominantly Anglo-Saxon is upheld by the dialect which we find in the county to-day.

We must turn next to the place-names of North-amptonshire. The termination *-by*, which, as before stated, shows the former existence of a Danish settlement, is of fairly common occurrence. We shall have occasion to refer to this in subsequent chapters dealing with the history of the county. Directly we begin to examine the names of the towns and villages of Northamptonshire, we cannot fail to be struck with the constant recurrence of forms which we know to be Anglo-Saxon. The three best known are *burgh* (borough) = city, *ham* = home, and *tune* or *ton* = town. Two of these can be seen combined in the name of the county town **Northampton**, which was originally (North) Hamtune. Borough occurs frequently, as for instance in Peterborough, "the city of Peter." The suffix *-ing* denoted "clan" or "family"; we continually find this in combination with *ton*, e.g. Nassington, or with *borough*, e.g. Wellingborough. In these cases the meaning is clear, but it is curious that in many instances (e.g. Thurning, Billing, Kettering, Wittering) we have the name of a family or clan applied alone to a place. *Stoke* (e.g. Stoke Doyle, Stoke Bruerne) and *stock* (e.g. Cotterstock and Brigstock) both mean stockade. *Thorp* = village, originally a Scandinavian word, occurs frequently (e.g. Apethorpe, Thorpe). The suffix *-ley*, as in Fawsley, means pasture, field, glade. The termination *-hoe* is the Anglo-Saxon word *hoh*, meaning a heel, a spur, a spur of a hill; Wadenhoe thus means Wade's hill. *Wick* means village, or hamlet, and Bulwick (spelt Bolenwyke in 1272) is "Bull's place." Fotheringhay is misspelt with an h; it should be Fotheringay, and represents the Anglo-Saxon *Fotheringa-ieg*, meaning "the island (or peninsula) of the Fotherings." The derivation of Oundle is generally wrongly given as Avon-dale; it appears in the *Anglo-Saxon Chronicle* as Undela, Undalum, dalum being the dative plural of *dœl*, "a dale." The Un-, in Professor W. W. Skeat's opinion, is short for Unnan, genitive of Unna, a man's name (on record). The meaning then of Oundle is "Unna's dales," and the reason for the name is apparent to anyone entering the town from the west.

The population of Northamptonshire by the 1911 census was 348,552 ; this includes the population of the Soke of Peterborough, 44,722. The valleys of the Nene and its tributaries are more densely peopled than the higher districts, the population of nine towns situated in these valleys amounting to half the total population of the county. There has been a considerable increase of population in such towns as Rushden, where thriving industries have attracted large numbers of people of late years.

10. Agriculture — Main Cultivations, Stock, Forests.

The agriculture of any district depends upon two principal things, firstly the climate which the district enjoys, and secondly the nature of the soil. Two other matters affect it considerably ; men are sure to try to grow those crops for which there is a demand, and therefore the needs of a district, or some densely populated parts near it, are bound to influence its agriculture. Again, the question of transport is sure to be important. If the means of sending away the produce of the land be good, an extra inducement is at hand to get the most out of the ground, but if on the contrary it is a difficult or expensive matter to send to other parts what the land yields, men are not likely to try to obtain more than is sufficient for local needs.

We have already learnt that the climate of North-amptonshire is on the whole an equable one, with a rainfall ranging from moderate in the west to low in the east. The nature of the soil varies with the locality; in the east in the fen district we find the black mould of the fens celebrated for its fertility; the uplands are for the most part covered with a brown crumbling loam, emi-nently suited for farming; elsewhere we find clays, or, as in the valley of the Nene, alluvial land. We may then expect to meet with all kinds of crops in the north-east corner and on the uplands, with wheat on the clays, and with permanent pasture in the floodlands of the Nene valley.

When we consider the question of demand, we must remember that wheat, if the conditions are suitable, is always worth growing (even its straw is of great value), that barley is needed for the important brewing industry of the Midlands, and that stock-raising is one of the most profitable occupations of the farmer. The last-named brings in its train the raising of such crops as oats, and various root and grass crops.

Finally there remains to be considered the question of transport. In a later chapter it will be seen that in this respect the county is singularly well served. It has no lack of good roads and waterways, although some of the latter are sadly neglected, and it is crossed by the main lines of four great railways, besides being covered with a network of branch lines. There can then be no difficulty with regard to transport.

It is hardly necessary to point out that success in farming depends upon an intelligent understanding of the main conditions of agriculture. One of the most important things that a farmer has to do is to find out what rotation suits his land best. If the same crop were to be grown upon a piece of land year after year, the soil would become so impoverished that it would be found impossible, without very heavy manuring, to make it yield a paying crop. Another reason for changing the crop from year to year is to discourage the attacks of the various plant and animal pests peculiar to each crop. For these reasons, the land is called upon to produce different crops in a more or less regular order, called a " rotation." Different soils of course need different treatment ; thus a heavy soil is found to do best with a rotation of beans, wheat, and bare fallow ; such heavy land is found on the boulder clay. With a lighter soil we find that the order that answers best is roots, barley, clover, and wheat. From a business point of view, the best rotation is always that one in which the best paying crop occurs most often without diminishing the productivity of the soil.

Year by year the Board of Agriculture sends out a full report on the agriculture of the whole country, and of each county in particular. If we obtain the reports for a succession of years, we can see how, in any particular county, some special cultivation has advanced or decreased in importance. There is generally some good reason for this ; the demand may have become greater or less, the weather in a particular year may have been exceptionally good or the reverse ; or, in the case of livestock, some disease may have been prevalent and caused great loss to the stock farmers.

Turning to the reports for Northamptonshire, the area of which is given as 638,612 acres, we find the following particulars for the years 1888, 1898, and 1908[1] :—

	1888 Acres	*1898* Acres	*1908* Acres	*Average* Acres
Corn Crops (Wheat, Barley, Oats, Rye, Beans, Peas)	145,000	127,000	117,000	130,000
Wheat (alone)	62,000	50,000	40,000	51,000
Green Crops (Potatoes, Turnips, Swedes, Mangolds, Carrots, Cabbage, Kohl-rabi, Rape, Vetches, etc.)	42,000	34,500	29,000	35,000
Clover, Sainfoin, and Grasses under rotation	35,000	37,000	32,000	35,000
Permanent Pasture	325,000	351,000	370,000	349,000
Bare Fallow, or Uncropped Arable Land	13,000	11,000	9,500	11,000

In the first place we must look at the relative importance of the different crops raised in the county, and for this purpose it would be well to take the average figures for the three years we have chosen. To begin with, we notice that the greater part of Northamptonshire (nearly 55 per cent.) is devoted to permanent pasture. This means that the rearing of live-stock for meat, hides, and wool is so important that the land of more than half the county is set aside for grazing. We know that there is a great demand for the commodities mentioned, and as the valleys of the Nene and its tributaries are well adapted for providing pasture for cattle, we can

[1] For the 1909 figures see the diagrams on pp. 214 and 215.

understand the important position occupied by live-stock.
We have alluded before to the fact that the Nene and its
tributaries are very liable to overflow their banks and
flood the lands bordering them. Every time that they do
this, the waters carry over the adjoining fields a fine silt
which they leave when the flood subsides, and thus the land
is continually manured without expense, making it possible
to keep it in permanent grass without the soil becoming
impoverished.

The next place of importance is occupied by corn crops,
which have on an average covered 130,000 acres (20 per
cent. of the county). Of these wheat comes first in
importance with an average of 51,000 acres (8 per cent.
of the county), and barley next with an average of 40,000
acres (6 per cent. of the county). These two crops differ
essentially in what they need for successful growing;
wheat wants a heavy soil, capable of holding moisture, and
a dry sunny summer; hence it is more grown in the south
and east of England than in any other part of our islands.
Northamptonshire possesses the stiff soil needed, and has
a low rainfall, so that it is well adapted for growing wheat.
For barley, on the other hand, a lighter soil is needed; its
roots do not penetrate deeply, and it needs a short hot
summer. In the lighter soils of its uplands Northamptonshire can therefore produce barley in great quantities. In
the northern part of the county some of the best malting
barley is grown, and Peterborough is one of the largest
barley markets. We may add that there was a notable
increase in wheat acreage in 1910; new methods,
labour-saving machinery, and the higher prices due to

greater world-wide demand have made English wheat-growing profitable. Another contributory cause has been the introduction of new French wheats.

A fairly large quantity of oats is grown, covering an average area of 24,000 acres (nearly 4 per cent. of the county). This is a hardy crop, intermediate to wheat and barley. It needs a cool, moist climate, and is therefore extensively grown in the more northern parts of the kingdom, but it plays an important part in feeding live-stock and is therefore grown to a fairly considerable extent in this county.

Very little rye is raised; it has a tall straw, however, which is of great use in thatching, padding harness, etc.

About equal areas, 35,000 acres (5½ per cent. of the county), are devoted to "Green Crops" and to "Clover, sainfoin, and grasses under rotation." The latter are for feeding stock and take their places in the rotation systems. Of the area given to green crops, more than one half is given to turnips and swedes, whose main use, like that of grasses, is as food for live-stock—more particularly in the winter. In the south of the county, where the soil is profitable for grazing, there are many dairy-farms which help to supply London.

We must next compare the figures for the three years which we are considering. Two facts stand out prominently; the area under permanent grass has risen steadily from just under 51 per cent. of the county to almost 58 per cent., while at the same time the acreage under corn crops has steadily decreased from about 23 per cent. to little more than 18 per cent. of the county. The falling off in wheat appears at first sight to account for most of this, but an examination of the figures for each crop shows that barley and peas also have decreased, that beans have hardly changed, but that oats have risen considerably. The drop in wheat-growing and the increase in the amount of land devoted to permanent pasture, taken in conjunction with the increased growing of oats, point clearly to the fact that the Northamptonshire farmer finds stock-raising more profitable at present than the growing of wheat. If we wish to increase the quantity of wheat produced in this country we must improve our methods of farming. Already we can raise from 30 to 50 bushels or more per acre, and it is hoped that with more scientific work upon the land we may be able to obtain a still greater yield.

We must turn now to the question of live-stock. From the same tables as those from which we obtained the figures for the principal crops of Northamptonshire we find that the live-stock returns for the county in the same years were as follows :—

| | England Average | Northamptonshire | | | Average |
		1888	1898	1908	
Horses	1,145,000	22,000	22,500	24,000	23,000
Cattle	4,675,000	124,000	122,000	136,000	127,000
Sheep	15,878,000	423,000	407,000	413,500	414,500
Pigs	2,179,000	30,000	29,000	36,000	32,000

The area of Northamptonshire is about 2 per cent. of that of England, and if we compare the average number of horses in the county with the number returned for all England, we find that the percentage is practically the

same. When, however, we consider the average number
of cattle in our shire, we find that the county has 2·7 per
cent. of the average number in England. Similarly we
find that Northamptonshire sheep amount to 2·6 per cent.
of those in England. The percentage of pigs is lower,
namely 1·5 per cent. Thus as regards horses we have
about the number that might be expected, and in the

Shire Mare, "Brown Duchess"
(*The property of Mr F. J. Steward, Brigstock*)

matter of pigs rather less; but in cattle and sheep we
have far more than the size of our county might lead us
to anticipate.

Again, if we compare the figures for the different
years, we notice a slight increase in the number of horses,
a greater increase in that of pigs, a small decrease in

sheep and a very decided increase in the number of cattle.
These results are in accordance with what we have learnt
from a study of the cultivations of the county.

Of the 24,000 horses in Northamptonshire in 1908,
17,000 were employed for agricultural purposes.   One of
the most important types is the shire horse, which comes
from a very old Lincolnshire and Cambridgeshire breed.

Shorthorn Bull, "Barrington Boy V"
(*Bred by Mr F. J. Steward, Brigstock*)

It is the largest of draught horses, and on account of its
strength it is specially suited for work in the stiff soils
encountered in these parts of England.   As regards
cattle, the shorthorn is the most widely distributed breed.
It is the type which has been found to answer best the
double need of the farmer—a beast which will give good

beef and yield good milk. The sheep are mostly of the Hampshire and Oxford down breeds, and in the north of the county the Lincolnshire and half-bred sheep.

Lastly we must turn to the woodlands of Northamptonshire. Three famous forests are found within its borders, those of Rockingham, Salcey, and Whittlebury. The first of these at one time extended over a great area along the north-west side of the county, but much of it has now disappeared. It is believed that the Romans cleared it in many parts, but the steady advance of agriculture and the use of wood as fuel must be held responsible for the greater part of the work of destruction. These woods have at all times been used chiefly for pleasure, for hunting in olden days and for the preservation of game in modern times, and little or no effort has been made to obtain wood for commercial purposes. Nevertheless Northamptonshire and Leicestershire have always been famous for their ash trees. At the present time there is a great dearth of ash, and as recently as 1899 the Coachbuilders' Association approached the President of the Board of Agriculture with the object of obtaining his help in promoting the cultivation of this tree in the Midland counties. It is said that rabbits are the great hindrance to this special cultivation ; no other tree is so liable to be peeled and damaged by them as ash. This necessitates such expensive wiring in setting out a plantation that the cultivation can hardly be made to pay.

The elm is very abundant, particularly varieties of the wych elm ; the beech, of which most beautiful specimens are to be seen in Lilford Park, is only found in cultivation. The woodlands chiefly contain oak, ash, and elm. The undergrowth is of blackthorn, maple, white-thorn, bramble, hazel, privet, spindlewood, dogwood, sallow, and—less common—buckthorn.

11. Industries and Manufactures.

It was pointed out in Chapter 2 that geographers often divide England into two by a line drawn from the mouth of the Severn to the Wash. North-west of this line lies industrial England, to the south-east of it agricultural England. Northamptonshire lies to the south of the line, but it possesses within its borders industries and manu-factures of great importance. Roughly speaking, the industries of the county are carried on in the central part, that is in the district between Northampton and Thrapston, and in Peterborough. The reasons for this ought to become apparent when we consider what these industries are.

First in importance come the leather trade and indus-tries allied to it, such as the making of boots and shoes. From early times the cattle on the rich grazing lands of the Nene valley supplied the hides, and the oak bark for tanning them was easily obtained from the neighbouring forests. If we remember the importance of the position of Northampton, lying close to the route of trade passing north and south, and commanding the Nene valley, we can understand how easily an important trade in leather and the commodities made from it might grow up. It is believed that in medieval England many towns and even villages tanned much of the leather needed for local requirements, but we have evidence that even in those days the leather trade of Northampton had more than local importance.

Tanning is the process by which raw hides and skins are turned into leather. The chief substance necessary for this is the bark of oak-trees, and it is by steeping the hides in the liquor produced from the bark that they are rendered fit for use. Another important process is that performed by the currier. Into his hands are placed skins such as those of calves, and by treating them with oil and fat he renders them fit for use for such purposes as the uppers of boots. Tanning by means of bark is a lengthy business ; indeed in the fifteenth century leather was not considered to have been thoroughly tanned unless it had been in the tanner's pit for a full year. During the last thirty years, however, a new process has been discovered— chrome-tanning. By means of salts of chromium the time necessary for converting raw hides into leather has been enormously reduced, and the material so obtained is stronger and better in every way than that obtained by the old processes. Fortunately Northampton was quick to see the advantages of the new method, and the local manufacturers of leather have held their own under the new conditions that have arisen. A large amount of leather is obtained from abroad, as the boot and shoe industry needs a far greater supply than can be produced locally, and uses many kinds of leather which cannot be produced at all in these islands.

It was not until about the middle of the seventeenth century that Northamptonshire began to specialise in the manufacture of boots and shoes. From that time onward the industry grew steadily, and the county became the most important boot-producing district in England. The year 1857 was a critical one, for then for the first time labour-saving machinery was introduced. Fortunately for the county the opposition to the change did not continue long enough to cause a lasting injury to the trade. Down the valley of the Nene from Northampton to Thrapston one cannot help noticing the succession of boot-factories. In town and village alike they are to be found, spoiling the quiet old-fashioned appearance of the countryside indeed, but reminding us that it is possible for great industries to be carried on in the midst of simple country surroundings, without the hurry and bustle and unhealthy conditions so often associated with a large town.

Apart from the introduction of machinery perhaps the most interesting development to be noted in this industry is the specialisation which has grown up in the manufacture of each part of a boot or shoe. Whereas originally a working bootmaker made the whole of a boot, now sixty or more different machines are employed for the making of a single boot, and each workman confines his attention to one process only of the many which are needed to produce the finished article.

The most important centres of the boot and shoe industry are Northampton, Wellingborough, Kettering, Irthlingborough, Higham Ferrers, Rushden, and Raunds, the last-named town specialising in army work.

Boots and shoes made in the county are sent to all parts of the world. At one time the manufacturers in

Northamptonshire had practically the whole of the export trade in their hands, but of late years a keen competition has arisen with the United States of America, whose manufacturers, by the employment of more economical methods, and by studying more carefully the wishes of their customers, have succeeded in gaining a considerable amount of the trade.   The enterprise of the manufacturers

"Pressing the Heel": a workroom in the Ocean Works, Northampton

of the county, however, has proved more than equal to this competition, and the export trade to the United States has recently made rapid strides.

In connection with leather we may mention two other industries.   In Northampton is to be found one of the oldest and most important firms of bookbinders in the kingdom ; and in Daventry a considerable manufacture

of whips was formerly carried on. This latter industry
has ceased now in Daventry, although it survives in
other towns of the county, but it is interesting to
notice that it flourished at Daventry when that town
was the centre of a great deal of road traffic, and died
away with the advent of railways and the consequent
disappearance of coaches.

From very early days there has been a considerable
trade in wool in Northamptonshire, and at different
periods cloth-making has been carried on, but the latter
has gradually died out, the industry naturally withdrawing
to other parts of our land where the conditions were more
favourable.

An industry which at the present time has reached a
very low ebb, but which it is hoped to revive, is that of
lace-making, the introduction of which craft seems to have
been due largely to French workmen, who after the Revo-
cation of the Edict of Nantes sought refuge in our country.
During the seventeenth and eighteenth centuries, and in
the early part of the nineteenth, lace-makers were to be
found in most of the villages of Northamptonshire. Mr
T. J. George, in an article on lace in the *Victoria County
History*, says—"It has been remarked that the hands of
Northamptonshire women are very small and well-shaped
compared with those of the women in other counties.
This is doubtless owing to the fact that so many of the
women of the previous generation worked at their pillows
from childhood, and their hands were not roughened by
coarse work or field labour." The industry has had
many ups and downs, but owing to various causes—chief

amongst them being the introduction of cheap laces made by machinery—it appeared doomed to die out. Quite recently, however, strong efforts have been made to revive it, and it is hoped that with proper encouragement the occupation of lace-making may once more be set on a firm footing in Northamptonshire.

Brewing is an important industry carried on in all the larger towns ; a great deal of barley is grown in the Midlands to meet the needs of the brewers, though for many reasons the quantity of beer consumed now is less in proportion to the population than in former days.

Amongst the smaller industries of the county we may mention brush-making, which is carried on in several places, especially at Northampton, Kettering, and Wellingborough. An interesting industry at Islip is the manufacture of horses' collars, the outer covering for these being made of rushes which are obtained from the valley of the Nene. There is a certain amount of basket-work carried on, and the osiers for this trade are obtained from beds situated on the banks of the river Nene. Paper-making has been carried on to a limited extent, and there are records of bell-founding, but this seems to have died out completely. At Finedon there used to be a considerable trade in dried apples, but this, too, has almost disappeared.

Of the mills situated on the river only five grind bakers' flour, one at Northampton (Nunn mills), the others at Houghton, Ditchford, Islip, and Water Newton. The rest either are "gristing mills" or are unused. One or two are employed for pumping and for electric light.

12. Minerals.

The two most important minerals worked in our
county are building-stone and iron. Northamptonshire
has been famous ever since the Roman period for
its building-stone ; while iron was almost certainly
smelted in the county from equally early times, wood
for this purpose being abundant, but the industry seems
to have ceased and only to have been revived in the
middle of the nineteenth century. We will turn our
attention first to the stone quarries, which are to be
found scattered over a large part of the county.

In the first place it should be observed that a stone
quarry in olden times could not hope to be of more than
local importance unless it was situated near a convenient
waterway. The methods of traction in use were too
costly as a rule to make it worth while to send building-
stone any great distance, unless the quarry from which it
was obtained was close to a navigable river. The famous
quarries of Barnack furnish us with an excellent instance.
Barnack lies between the Welland and the Nene, at the
side of a narrow low strip of country between these rivers,
at one of the two places where they approach nearest to
each other. We have ample evidence that the stone of
Barnack was carried far afield in early days. Religious
houses, scattered over the counties in the east of England,
possessed quarrying rights there, and Peterborough in
particular derived immense benefits from having so fine a
supply of building-stone almost at its gates. The quarries

however were worked out long ago—it is said that they were exhausted by the end of the fifteenth century—and now all that is to be seen of these famous stone-pits is the grassy tract of land known as " Hills and Holes."

The stone obtained from these quarries formed a part of the Lincolnshire Oolite series ; it was a coarse free-

Stone Quarry at Weldon

(*Showing peculiar saw*)

stone, formed for the most part of minute shells or pieces of shell, bound together by a matrix of the same chemical substance.

Another place famous for its quarries is Weldon. Like Barnack, Weldon has been renowned for its stone for many hundreds of years, but, unlike Barnack, quarrying at Weldon still flourishes. The pits at the latter place indeed now yield the most noted building-stone of Northamptonshire. Weldon is not so fortunately situated as Barnack, as it is much farther away from both the Welland and the Nene, but it now enjoys greater facilities than formerly for sending its stone away, as in 1880 the Midland Railway opened a branch line from Kettering to Manton which passes close to Weldon. It is claimed for this freestone that it is more durable than most of the oolitic limestones used for building ; its power to resist the attacks of weathering agents being due to the peculiarly strong way in which nature has cemented together the minute shell particles which form the basis of the stone. The supply of freestone still unquarried is reported to be practically inexhaustible. It is obtained in the simplest manner, being sawn down in great rectangular blocks until the " bed " is reached, when wedges are driven in beneath. These loosen the entire block, which is then placed upon a trolley and hauled up ready for carting.

Many modern churches and college buildings at Oxford and Cambridge have been constructed of Weldon stone. In 1890 the pinnacles of Rochester Cathedral were restored with it ; and the same stone was used for refacing the whole of the central tower in 1904.

The quarries in the middle and south of the county have for a great length of time provided large quantities of building-stone, mainly however for local purposes. Some indeed, as at Cosgrove and Stratford, have been of importance, and the district round Northampton has

yielded huge quantities of this building material, but the
finer types of stone are to be found in the north.   It

Splitting Colly Weston Slates

is interesting in a journey up the valley of the Nene to
notice that the stone used for building tends to show
tints of yellow or red in a more and more marked degree

as we approach the south-west end of the county. This colouring is due to the presence of iron oxide in the stone, and in some districts, such as Kettering and Northampton, ironstone has been much used as a building material.

At Colly Weston, Easton, and Kirby stone "slates" are quarried. Large pieces of stone are obtained from the pits, and are exposed to the action of frost; they are even watered in the evening in order that the frost may split them the more easily. After they have been exposed in this way a single tap from a hammer is sufficient to split the stone into thin slabs, which are eminently useful for roofing. Colly Weston slates are famous throughout the Midlands; they form a very substantial but heavy roof, as, besides being of great weight themselves, it is as a rule necessary to cement them as well as fasten them with pegs. One of the most typical sights of a north Northamptonshire village is a row of cottages with Colly Weston slated roofs, the rafters of which have warped and bent under the great load they have had to bear for a hundred years or more.

The production of iron is an industry which has been increasing steadily in Northamptonshire in recent years. For this industry three things are necessary—the ore, limestone for a flux, and coal for smelting. The ore is found in large quantities between Northampton and Stamford, and in lesser quantities elsewhere; limestone is plentiful, and coal can be obtained easily by rail. The ore occurs in the bed of rock known as "Northampton Sand"; it is hydrated ferric oxide, and is called brown hematite. The beds vary considerably in thickness;

Ironworks at Corby

they thin out and disappear south of Northampton, but
the average thickness of those worked is about ten feet.
The way in which the ore is obtained is as follows;
the soil and beds of rock which overlie it are removed
and the ironstone is then quarried out. Into the holes

Running off Slag: Islip Ironworks

thus formed the original top layers are thrown and the
whole roughly levelled, so that after an interval of a
few years agriculture is once more possible in these
places. The ore is reduced in huge blast furnaces, from
which the molten iron is drawn off at regular intervals
into sand moulds; the by-product—the dross, as it were

—is the slag, which is drawn off into large vessels ready to receive it, and, when it has cooled sufficiently, it is broken up and used as road-metal. This slag is porous, and experiments are being made at present with the object of producing from it a more satisfactory material by forcing tar into it, and thus rendering it impervious to wet, and therefore less liable to be broken up by the action of frost.

Bricks are made in many parts of the county, as the necessary clay is found in abundance, but the increase in competition and the introduction of improved labour-saving machinery have caused most of the smaller brick-yards to close. The greatest brickmaking district is that which lies south of Fletton, close to Peterborough, but it is outside Northamptonshire.

Lime is obtained from the Great Oolite limestone, and as it can be prepared from other limestones found within the county, it is generally procurable at any place in Northamptonshire.

Portland cement is made at Irthlingborough from Great Oolite limestone and Upper Lias clay. At Finedon "flags" are made for paving from a mixture of Portland cement with granite chippings and dust.

The industry of pipe-making has died out; it arose originally from the possession in Northampton Field of a clay especially suitable for the purpose. The supply of clay gave out some time before 1850, and after that date the clay had to be obtained from Devonshire.

13. History.

As the boundaries of our county are largely artificial, and the term Northamptonshire does not stand for a district marked off from the rest of England by distinctive features, it cannot be said to possess a marked history of its own.   All that can be done is to set down whatever of historical interest has happened within its borders, and to remember that the factors of importance with regard to the county in connection with its history are (1) its central position, and (2) the fact that it is crossed at each end by a great road running north from London, and that these roads are themselves connected by the Nene valley.

We may begin with the Roman era, where, indeed, history properly so-called begins.   We must remember of course that in the time of the Romans there were no such divisions of the land as our modern counties. The tract of land which we call Northamptonshire did not exist as a unit, but it is interesting to see what marks were left here by the Roman occupation of Britain.   As we have just pointed out, our county is crossed by the two great routes leading northwards from London, which, roughly speaking, follow the same directions as those taken nearly 2000 years ago by two great Roman roads.   One, called Watling Street, ran north-west from London to Chester, passing through the west end of Northamptonshire by Towcester.   The other, known as Ermine Street, ran north from London to York, passing through the east end of our county by Castor.

The names of the towns which we have just mentioned
show clearly their Roman origin ; Castor from the Latin
castra, a camp ; Towcester from the river Tove on
which it stands, and the Latin *castra*. The word *castra*
occurs in several forms in the names of our towns to-day.
In Northamptonshire, besides the two mentioned above,
we have Irchester ; in other parts of England many other
forms occur. It is worth noticing that neither of the two
most important towns of to-day, Northampton and Peterborough, lay upon these roads.

From the Roman remains already examined in our
county we can make the following assertions. There
was one considerable town Castor (not wholly within
Northamptonshire, for part of it lay south of the Nene in
Huntingdon), and three small towns, one at Irchester,
another at Towcester, and the third somewhere between
Whilton and Norton on Watling Street. Scattered over the
county were villas and country dwellings, e.g. at Apethorpe ;
and there were two industries of some moment, namely,
ironstone diggings, wood from the forests being used for
smelting the iron, and a large pottery manufacture carried
on at Castor, where the necessary clay could be obtained.

The Roman occupation of Britain came to an end
about 410 A.D., and we then reach the Anglo-Saxon
period. The Saxon settlements were made earlier than
those of the Angles, but the latter supplied the greater
part of the new population which settled over England.
The Saxons appear to have entered Northamptonshire
from the south, moving northwards from the Thames ;
the Angles entered the county from the east, moving up

the valley of the Nene, and from the main river they struck off up the tributaries, following them to their sources. Many interesting remains have been discovered of these early settlements, chiefly from burial grounds which have been examined. From these it appears that cremation was common before the introduction of Christian burial. Two important facts stand out clearly; firstly that the new races avoided settling on or near the old Roman roads, which thus served as dividing lines between batches of settlers; and secondly that they almost invariably chose as sites for their towns or villages places close to natural springs, or where only shallow wells were necessary to obtain water. Such spots were found in plenty at the junction of the Northampton Sand with the Upper Lias clay.

It is interesting to notice how many towns and villages bear names which show their Anglo-Saxon origin. When the land came to be divided up into kingdoms, Northamptonshire formed part of Mercia, which had spread southwards under the pagan king Penda. In the ninth century a large part of central England, including our county, was overrun by the Danes. By the treaty made between Alfred and Guthrum in 878 A.D. most of Northamptonshire seems to have been incorporated in the Danelagh, but the district was reconquered by Edward the Elder with the help of his sister, who was known as "the Lady of Mercia." Traces of Danish settlement are not very numerous, but, as we have seen, the termination -*by* in the name of a place shows that it was settled by the Danes. Several names with this ending are found in the north-west part of the county (e.g. Cold Ashby, Naseby, Thornby, Kilsby, Barby). Canons Ashby, it may be added, is said to be the most southern place in England whose name ends in -*by*.

We come next to the Norman period. The forests of Northamptonshire seem to have proved a great attraction to the early Norman kings, who were great hunters. In those days Rockingham Forest extended over a large area, and there were in addition the forests of Salcey and Whittlebury in the south. The castles of Rockingham and Northampton, both built soon after the Conquest, were frequently visited by the kings, and in them several famous historical scenes took place.

In 1290 Edward I's wife, Queen Eleanor, died at Harby, and her body was removed to Westminster for burial. At every place where it rested for the night a beautiful memorial cross was afterwards erected. Only three of these remain now, two of them being within our county, one at Northampton, the other at Geddington.

Four parliaments have been held at Northampton. The first was held in October, 1307, immediately after the death of Edward I, when such matters as the coronation and marriage of Edward II were considered. In 1328 the First Statute of Northampton was passed by a parliament which sat at Northampton for three weeks at Easter. It met again in the same town in July, 1338, to prepare for the war with France, but it broke up suddenly as the King was forced to march north to meet the Scots. In 1380 the Second Statute of Northampton was passed by the last parliament which met at

Queen Eleanor Cross, Geddington

Northampton. It also passed the Poll Tax, one of the chief causes which led to the rebellion of Wat Tyler and Jack Straw.

Fotheringhay Castle, of which little remains to be seen but the keep-mound, was the scene of the execution of Mary Queen of Scots in 1587. Her body was removed for burial to Peterborough Cathedral, where it remained until 1612, when it was taken to Westminster.

In the Civil War Northamptonshire, owing to its central position, was the scene of much fighting. Many events of military importance took place round Daventry, where Charles I fixed his headquarters shortly before the disastrous battle of Naseby in 1645. In Holdenby House the King was kept virtually a prisoner from early in 1647 until he was taken south by Cornet Joyce.

No historical events worthy of record occurred in our county in the eighteenth and nineteenth centuries. As far as Northamptonshire has been concerned the last two centuries have witnessed nothing but peaceful progress in manufactures and agriculture. The manufacture of boots and shoes has come to be an industry of the greatest importance ; the production of iron has been increasing steadily for years and will in all probability continue to increase in importance. We may notice also that much has been done to check the recurrence of the serious floods to which the valley of the Nene is subject.

14. History of Northampton and Peter= borough.

We will now turn for a moment to the history of the two great towns of our shire.

(1)   *Northampton.*

One of the most important things to note in connection with any town is its position. Taking Northampton
first, there are four points to be specially remarked with
regard to its situation.   It lies near a great route leading
northwards, midway between the ancient capitals of
England, Winchester and York; secondly, it is at the
head of the Nene valley; thirdly, it is situated on a
ridge of high ground overlooking the Nene; and lastly,
it is in the centre of a great agricultural district.   Hence
it occupies a typical position for an important inland
town.

Its origin is obscure; all that we can say for certain
is that it was of sufficient importance in Anglo-Saxon
times to give its name to the county in which it was
situated when the land was divided up into shires.
With its name we dealt in the first chapter.   We
know that the town was occupied by the Danes for a
few years during the tenth century, and that it was
ravaged by them in 1010.

The first great name in the annals of the town is
that of Simon de St Liz (or Senlis).   He was a knight
who came over to England with the Conqueror, and
married a daughter of Earl Waltheof, who brought to

him the earldoms of Northampton and Huntingdon. He
appears to have done a great deal for the chief town of
his earldom. He rebuilt it and fortified it, he reared
the castle on the hill, and he endowed the Priory of
St Andrew as well as many churches.

The history of the town was for a long time bound
up with that of its castle, which we have traced in the
chapter devoted to Military Architecture. Our early
English kings frequently paid visits to the castle, attracted
chiefly by the fine hunting which was to be had in the
neighbouring forests of Rockingham and Whittlebury.
One historic event, deserving of description, occurred
within its walls. In 1164 Henry II held a great council
in his castle at Northampton, and thither he summoned
Becket, Archbishop of Canterbury, to render an account
of all the money which the Archbishop, when chancellor,
had received from the king. Becket attended, but refused
to accept the decision of the council, and after a stormy
scene in the Great Hall he appealed to the Pope. The
same night he made his way out of the town, and escaped
to France.

The general importance of the town in these early
days is attested by the number of councils held within
its walls, by the frequency of the visits paid by sovereigns,
and by the privileges granted to the township by succes-
sive kings. The first charter was given by Richard I in
1189, and John appears to have visited the town very
regularly; indeed Shakespeare's play of *King John* opens
with a scene at Northampton. About the middle of the
thirteenth century, between 1230 and 1258, a University

seemed destined to become established in the town, owing to quarrels at Oxford having led to the removal of many of the scholars to Northampton, but the dispute was settled and the scholars returned to Oxford. In connection with the central position of Northampton it is interesting to find records of several bands of crusaders having made that town their gathering centre before setting out for the Holy Land. In 1240, for example, a band of men under the Earl of Cornwall assembled here before they started for Jerusalem; and in 1267 Prince Edward and a hundred knights assumed the crusaders' badge here, in the presence of the King and Queen. The form of St Sepulchre's church, in imitation of that of the church of the Holy Sepulchre at Jerusalem, is of especial interest in view of these gatherings of crusaders taking place at Northampton.

In 1290 Edward I lost his consort, Eleanor of Castile; the memorial cross at Northampton, erected in her memory, still remains.

The town grew steadily in importance, as may be judged from the fact that it sent two burgesses to the parliament summoned by Edward I in 1295. Within its own walls several parliaments were held, to which allusion has been made in the preceding chapter.

In 1460 a battle took place outside the town between the forces of Henry VI and those of the Earls of Salisbury, Warwick, and March, in which the King was defeated and taken prisoner. In the Civil War Northampton sided with the Parliament against the King, and several memorable fights took place within a short distance of the town.

Interior of St Sepulchre's Church, Northampton

From a political point of view the importance of
Northampton may be said to have decayed with its
castle; its trade however and its commercial importance
grew steadily. One by one it gained charters of liberties,
and gradually freed itself from the interference of outside
authorities. We have already mentioned thàt Richard I
granted the first charter to the town[1].

In the official guide-book, edited by Mr C. A.
Markham, issued for the Church Congress held at
Northampton in 1902, many interesting facts relating
to the history of the borough are gathered together. It
tells how, "in 1299 the burgesses obtained, by the
king's charter, the right to choose their own mayor,
and two bailiffs every year. This was another great
step towards self-government. In 1301 the king made
an extensive grant of tolls for rebuilding the walls, and
it is probable that at this time the area of the town
was enlarged, and the centre of the borough transferred
from the Mayorhold to the Market Square, where the
Town Hall was erected at the corner of Abington
Street." "In 1489 an Act was passed for regulating
the election of the mayor and the eight and forty; and
placing the government of the town in the hands of the
mayor, ex-mayors, bailiffs, ex-bailiffs, and forty-eight
burgesses. The authorities of the town continued to be
appointed under this Act until the Municipal Corporations
Act of 1835."

Guilds were formed amongst the craftsmen, of which

[1] Details of this and others will be found in *The Records of the Borough
of Northampton*, by C. A. Markham and J. C. Cox.

Queen Eleanor Cross, Northampton

that of the shoemakers has always been the most important. The trade in boots and shoes has taken precedence
of all others ever since the reign of Edward VI, and records
exist of contracts to supply boots to armies in the field as
far back as Cromwell's time. Leather, of course, has for
centuries been one of the chief commodities dealt in, and
great quantities of this used to be sold at the fairs held in
the town. The leather used now comes from all parts
of the world, but originally the supply was due to the
fine grazing grounds of the county and the ease with
which oak-bark could be obtained from the neighbouring
forests. In 1675 occurred a great fire which swept away
most of the old buildings, but from what we know of our
old towns, what appeared at the time to be a disaster was
probably a blessing in disguise. The borough has made
very great progress during the past century; its population
has increased from 21,000 in 1841 to 41,000 in 1871,
76,000 in 1901, and 90,076 in 1911, a growth due
mainly to the facilities afforded for communication and
trade generally by the construction of railways, and also
to the enormous development of the boot trade.

(2) *Peterborough.*

Peterborough is situated at the edge of the Fens, at
the point where the valley of the Nene opens on to the
flat plain through which the river flows to the sea. It lies
moreover near the great route running north from London
by way of the east coast. It has always formed a convenient centre for the produce of a great part of the fen
district, and on it all tracks across and along the Fens

converged.   Unlike Ely it is not built on an "island,"
but on the edge of the marsh country.

There is nothing to lead us to believe that Peterborough had its origin on an old British or Roman site.
On the contrary, we have in it an interesting example of
a town which grew up round a great monastic establish-

Bridge over the Nene, Peterborough

ment.   In 655 A.D. a Saxon priest, called Saxulf, founded
a monastery here.   As was usual, to supply the requirements of the monastery, a village rose round it, which
was known at first as Medeshamstede (homestead of the
meadows).   The monastic buildings were utterly destroyed
by the Danes in 870.   Soon after this we find the
town known as Burgh, and later as Peterborough.   The

Peterborough Cathedral and Bishop's Palace

rebuilding of the monastery was completed shortly after the year 970 by Ethelwold, bishop of Winchester, and remains of this building were discovered during some excavations which were necessary between the years 1883 and 1893. The monastery passed through troublous times in the days of the Conquest; it was laid waste by Hereward "the Wake" (who spared the church), and it suffered from two serious fires. When we remember that wood was the chief building material in those days, we can understand that the town suffered considerably from these fires. Finally in 1118 the foundations were laid of the great Abbey Church which slowly developed into the cathedral we now see. Originally the town with the rest of the county was in the diocese of Dorchester and afterwards of Lincoln, but at the Dissolution a new diocese was created, and the monastic church became the cathedral of the see of Peterborough, which included the counties of Northampton, Leicester, and Rutland.

We have already pointed out that the village known as Medeshamstede grew up round the spot where the monastery was first established. This naturally increased in size and importance as the monastic establishment grew greater, and in time it became an important agricultural centre. By a charter granted to it in the reign of Edward IV it returned two members to Parliament, and this right it exercised until 1885, when the "Redistribution of Seats" Act reduced the number to one. The town was the head of the Liberty or Soke (an Anglo-Saxon word meaning a place privileged to hold local courts) of Peterborough, a district co-extensive with the hundred

Peterborough Cathedral, West Front

of Nassaburgh. The rights of the Soke to-day are referred to in the chapter on Administration and Divisions.

It was not until nearly the middle of the nineteenth
century that the city began to grow to any considerable
size. Up to that time it had been nothing but an ecclesiastical and agricultural centre. From 1840 onwards,
however, when the railways of England were being
constructed, the population grew by leaps and bounds.
In 1841 the city contained only 7,000 inhabitants; in
1871 this number had increased to 15,500, in 1901
to 32,000, and in 1911 to 33,578. The causes of this
increase are not difficult to explain; the city became an
important railway junction, with communication north
and south by the Great Northern Railway, east by
the Great Eastern, west by the London and North
Western and the Midland. These improved means
of communication gave a great impetus to its trade
in agricultural products, live-stock, and timber. The
railways, moreover, provided a great deal of employment
by the erection of works and sheds, those of the Great
Northern at New England and of the Midland at Spital
being the most important. One more source of increased
trade is to be found in the great brick-making works
situated in the district.

Up to the year 1874 the government of the city had
been in the hands of the lord paramount, the Custos Rotu-
lorum (keeper of the rolls), and magistrates appointed by
the Crown[1]; but in that year a royal charter was obtained

---

[1] As part of the Soke; the Court of Quarter Sessions of which had
most extensive powers.

The Cathedral Close, Peterborough

by which the city was incorporated as a borough, with a mayor, aldermen, and councillors, and since that time it has been under municipal government.

15. Antiquities.

In Chapter 9, which dealt with the people and races of the county of Northampton, we pointed out that it is usual to distinguish four "Ages" before the Romans came to Britain.   These "Ages" can only be traced by the remains belonging to each which we have been able to identify, and in this chapter we propose to consider briefly these ancient relics, which we tell us a great deal about the people who inhabited this land so many centuries ago.

Of the earliest of these four pre-Roman periods, the Palaeolithic or Old Stone Age, we have discovered at present very few traces in our county.   In gravels found in the Nene valley, washed down from higher ground ages ago, roughly-chipped stone implements have been found, which certainly belong to that age.   Two of these were in the possession of the late Sir John Evans; one came from gravel obtained near Oundle, the other from a ballast pit at Fotheringhay.   Recent discoveries have considerably added to this list.

There are more remains, however, of the Neolithic or New Stone Age in our county.   One of the typical features of this period is the long barrow, pointing east and west, in which men of the Neolithic Age buried their dead in a crouching attitude, often interring with them their weapons. These, although still made of stone, were a great advance on those in use in the preceding age,

Palaeolithic Implements, found at Woodstone, near Peterborough

being indeed often polished and ground to an edge as we have already stated, and of a more finished description. Mr T. J. George, in the *Victoria County History of Northamptonshire*, gives the following list of Neolithic remains found within the borders of the county; they are for the most part polished celts or axes.

"Five good specimens have been obtained from Northampton; portions of four celts, with one perfect

Neolithic Javelin and Arrow-heads, found at Duston

one of a peculiar green slaty kind of stone, were found between Gretton and Kirby Hall; other specimens have come to hand from King's Sutton, Everdon, Towcester, Courteenhall, Great Harrowden, Weldon, King's Cliffe,

Castor and Eye near Peterborough. Flint arrow-heads have been found at Duston, and Oundle (Lyveden, Harrington, Brixworth, Dallington and Hardingstone), and a finely worked spear-head or dagger of flint was obtained at Norton by Mr B. Botfield in 1862...Hammer-heads of stone have been found at Singlesole in the Fens and from the neighbourhood of Gretton. Worked flints, such as the so-called thumb-flints, have been obtained from Borough Hill, Hunsbury Hill, Blisworth, Roade, Moulton, etc. Burials of this period have been noted at Great Houghton and Norton; with the remains at the latter place was found an earthen vessel as well as the spear-head mentioned above."

Since this article was written, discoveries have been made which warrant us in claiming a Neolithic site in the parish of Duston. Some 12,000 to 15,000 pieces of worked flint have been found there; they comprise between two and three thousand scrapers, about the same number of flakes, and about 125 arrow-heads, fabricators, borers, saws, cores, hammer-stones and indefinite implements, all of which are now to be seen in the Museum at Northampton.

Our county is not rich in remains of the next period, the Bronze Age. In the Museum of the county town are to be seen some cinerary urns belonging to this period which were discovered in 1890 at a spot between Great Weldon and Corby. With them were found a skeleton in a sitting position, and a bronze weapon. Other urns have been found at Brixworth, Desborough, Cransley, and Rothwell. In a barrow near Oundle two vessels of the period were found, and are now in the British Museum. Not many implements or weapons of bronze have been brought to light in Northamptonshire.

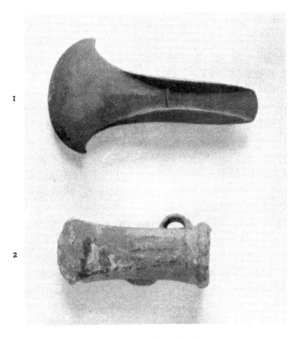

1. Bronze Palstave found at Aynho
2. Bronze Socketed Celt, found near Northampton

A sword, some palstaves, socketed celts, axes, and spear-heads have been discovered in various places.

In remains of the next period, the pre-historic Iron Age, Northamptonshire is far richer. At the end of a ridge of high ground about two miles south-west of Northampton stands Hunsbury Camp. It used to be thought that this was the work of the Danes, but we know now, from the excavations that have taken place, that it is certainly pre-Roman. The camp is oval in shape, covering an area of about 4 acres, and was fortified by a ditch 50 to 60 feet wide and about 15 feet deep. Here extensive digging was undertaken in 1882 in order to reach the ironstone which lay beneath the surface. The men engaged in the work discovered more than 300 pits, between 5 and 6 feet in diameter, some of which had been sunk as far as the ironstone. These pits were filled with soil, and in them were found many articles belonging to the pre-historic Iron Age. Amongst the implements of iron were found knives (still in their horn handles), spear-heads, swords and sword-scabbards, saws, nails, adzes, sickles, and other articles, the use of some of which is uncertain. In addition many pieces of pottery, some of which bear characteristic Late Celtic decoration, were unearthed, nearly all intended for domestic purposes, besides millstones for grinding corn, spindles, whorls for weaving, bone combs for carding wool, and the bones of men and animals. We may add that coins belonging to this period, some uninscribed and some inscribed, have been found in various parts of the county.

Other earthworks within Northamptonshire are to be found at Charlton, Daventry, Arbury Hill near Thenford, and Farthingstone. Close to Charlton stands Rains-borough Camp, oval in shape, and covering about 6 acres. There are two ramparts with a deep ditch between. The general view is that this camp is pre-Roman, but Roman coins have been found there, and it is probable that the Romans used the camp. To the east of Daventry, and overlooking it, is Borough Hill, where there is an old earthwork, probably pre-Roman, but certainly made use of during the Roman occupation of our island. The camp was probably one of the largest in England; its shape was roughly that of a parallelogram, the whole distance round being about 2¼ miles, and the area 150 acres. Many interesting discoveries have been made in it; remains of Roman buildings, and urns and pottery, some of which belong to a date earlier than Roman times. Near Thenford, at the top of Arbury Hill, and at Farthingstone there are earthworks which are believed by some authorities to be of pre-Roman origin, but at present there is so little evidence with regard to their age that we cannot be certain as to the period to which they should be assigned.

We come next to the Roman period. It ought not to surprise us that the Romans have left many marks of their occupation on Northamptonshire, when we remember that two of their chief roads, Watling Street and Ermine Street, crossed it, one at each end. There is some evidence for believing that two minor roads joined Watling Street, one near Towcester and the other at Norton, but of these we cannot be sure. Ermine Street threw off a branch at Castor known as King Street, which led due north to Lincoln, and there were possibly two other branches near Castor, one leading to King's Cliffe and the other to Peterborough. Two more cases of possible

Roman roads may be cited—one from Medbourne in Leicestershire to Stanion, the other from Tichmarsh in an easterly direction to join Ermine Street. It should be noted that the Romans made no road in this county to join Watling Street and Ermine Street; communication

Pottery made at Castor, found at Duston

between the two was made easy by the Nene valley. The most striking feature of all these roads is their straightness; the method of construction varied with the type of country traversed and the materials available, but the thoroughness with which they were constructed is always noteworthy.

We can now turn to the traces of Roman dwellings in our county. The remains discovered at Castor are so extensive that without doubt a considerable Roman town once existed there. The foundations of the buildings are still in existence, and it is possible to distinguish houses of different types, as well as pottery works. The district has not yet been thoroughly excavated, but from work that has already been done it has been found possible to reproduce the ground plans of some of these buildings. Many interesting things have been found in the district, including some beautiful mosaic pavements. At Irchester there are sufficient indications that a Romano-British village once existed there. Pieces of sculpture, a tombstone, ware of various kinds, bronze vessels, tools, etc. have been dug up, and it has been found possible to trace out roughly the shape of the "camp," which was surrounded by a stone wall 8 feet thick; but the site has not yet been thoroughly explored, and we cannot hope to know more about its history until further excavations have been carried out. Towcester has been identified with the Roman station Lactodorum on Watling Street. Sufficient remains have been found to satisfy us that it was once a Roman site; ramparts have been traced which probably once formed the boundaries of the settlement, and the foundations of buildings have been met with, but no very interesting relics of Roman occupation have so far been discovered.

Scattered over Northamptonshire traces have been found of many Roman villas or country houses. A study of a map on which all spots with Roman remains have been marked, shows clearly that the evidence collected

up to now points to the following districts as having been
most favoured in Roman times :—Ermine Street and
the adjoining country, Watling Street and the land bordering it (especially on the Northampton side), and the
valley of the Nene. We have not space to do more than
mention a few of the most interesting discoveries; an important villa at Apethorpe, mosaics at Cotterstock, a villa
at Great Weldon, Roman bricks in the walls of the Saxon
church at Brixworth, villas at Daventry and Whittlebury,
and a great collection of pottery and coins at Duston.
At the last-named place a silver bowl has been found
which is believed to have formed part of a water-
clock.

It is generally thought that the iron of the county
was worked in Roman times, but hitherto no conclusive
evidence of this has been forthcoming. Of the pottery
industry at Castor, however, we have abundant proofs.
Kilns in which the baking was done have been excavated,
and a great many specimens of the pottery made there
have been dug up. The characteristic "Castor ware,"
according to Mr F. J. Haverfield, appears to have had
"a white or whitish paste, coloured outwardly a dull slate
colour, blue or coppery in tint."

Many burial places of Anglo-Saxon days have been
discovered, and in the graves many different kinds of
ornaments and weapons have been found. It appears
that cremation and the burial of bodies in a north and
south direction gave way in the second half of the seventh
century and the first half of the eighth to burial in an
east and west direction—a change due to the influence

of Christianity, which in those years gained a firm hold on this part of England. Other interesting relics that we have of Anglo-Saxon times are the towers of the churches at Barnack, Earl's Barton, and Brigstock, and parts of the church at Wittering.

16. Architecture—(a) Ecclesiastical.

Northamptonshire has long been famous for its churches. When we remember the magnificent supply of building stone that the county possesses, the facilities it has enjoyed for water-carriage, and the influence which Peterborough must have exerted, we need not be surprised to find that no county in England excels it in its churches. In this chapter we propose to state briefly the chief characteristics of the various architectural periods, and to point out where examples of the work done in them may be seen in the county.

In the early days of Christianity no buildings existed which were suitable for congregational worship. The first structure which is supposed to have been adapted for Christian worship was the Basilica of Roman times, which was merely a hall, rectangular in shape, and divided into three portions by two rows of pillars—the foreshadowing of the nave and aisles of our churches. Changes of course took place, the most notable being the addition of north and south wings, the origin of our transepts. It used to be thought that at Brixworth in our county we had the remains of such a building; but this view is no longer held. All that we can be sure of is that there was originally at this place a considerable Roman building, some of the materials of which were used in the construction of the church.

The greater part of the buildings in the early centuries of the Christian era seem to have been composed of wood; it was not until the eleventh century that building in stone became general, and it is from this time that we date the first of the great building periods.

A preliminary word on the various styles of English architecture is necessary before we consider the churches and other important buildings of our county.

Pre-Norman or—as it is usually, though with no great certainty termed—Saxon building in England, was the work of early craftsmen with an imperfect knowledge of stone construction, who commonly used rough rubble walls, no buttresses, small semicircular or triangular arches, and square towers with what is termed "long-and-short work" at the quoins or corners. It survives almost solely in portions of small churches.

The Norman Conquest started a widespread building of massive churches and castles in the continental style called Romanesque, which in England has got the name of "Norman." They had walls of great thickness, semicircular vaults, round-headed doors and windows, and massive square towers.

From 1150 to 1200 the building became lighter, the arches pointed, and there was perfected the science of vaulting, by which the weight is brought upon piers and buttresses. This method of building, the "Gothic,"

Early English Lancet window, Nassington

Geometrical window, Oundle

Decorated window, Cotterstock

Perpendicular window, Lowick

8—2

originated from the endeavour to cover the widest and loftiest areas with the greatest economy of stone. The first English Gothic, called "Early English," from about 1180 to 1250, is characterised by slender piers (commonly of marble), lofty pointed vaults, and long, narrow, lancet-headed windows. After 1250 the windows became broader, divided up, and ornamented by patterns of tracery, while in the vault the ribs were multiplied. The greatest elegance of English Gothic was reached from 1260 to 1290, at which date English sculpture was at its highest, and art in painting, coloured glass making, and general craftsmanship at its zenith.

After 1300 the structure of stone buildings began to be overlaid with ornament, the window tracery and vault ribs were of intricate patterns, the pinnacles and spires loaded with crocket and ornament. This later style is known as "Decorated," and came to an end with the Black Death, which stopped all building for a time.

With the changed conditions of life the type of building changed. With curious uniformity and quick-ness the style called "Perpendicular"—which is unknown abroad—developed after 1360 in all parts of England and lasted with scarcely any change up to 1520. As its name implies, it is characterised by the perpendicular arrange-ment of the tracery and panels on walls and in windows, and it is also distinguished by the flattened arches and the square arrangement of the mouldings over them, by the elaborate vault-traceries (especially fan-vaulting), and by the use of flat roofs and towers without spires.

The medieval styles in England ended with the

Earl's Barton Church

(Showing Saxon tower: the battlements are added)

dissolution of the monasteries (1530–1540), for the Reformation checked the building of churches. There succeeded the building of manor-houses, in which the style called "Tudor" arose—distinguished by flat-headed windows, level ceilings, and panelled rooms. The ornaments of classic style were introduced under the influences of Renaissance sculpture and distinguish the "Jacobean" style, so called after James I. About this time the professional architect arose. Hitherto, building had been entirely in the hands of the builder and the craftsman.

We can now turn to the characteristics of each of these periods, and point out some of the best examples of each which are to be found in Northamptonshire.

Saxon.

The most noticeable feature in the Saxon church was its tower. This was built without buttresses, and either remained of the same size from the ground to the top, or else grew smaller by stages. Good examples of such towers are to be found at Barnack, Earl's Barton, and Brigstock. We have already pointed out that in the years before the period which we call Saxon, the most common building material was wood, and in buildings of the Saxon period it is interesting to observe a method of using stone which is known as "long-and-short" work. This was the practice of laying stones with the long side alternately horizontal and vertical, which, combined with the custom of building into the walls courses of projecting stone, reminds us of wood-work, and it is said that the method was adopted in order to bind the building

Wittering Church

Brigstock Church, showing Saxon Towers

together in imitation of the method used with timber. This " long-and-short " work can be seen in the towers of Barnack and Earl's Barton and in the corners of the chancel and nave of Wittering. The church at Brigstock also has a Saxon tower; this consists of a round tower attached to a square one of the same date. One more characteristic of the period may be mentioned; windows and doors were either angular-headed or round-headed ; in the former case the head was formed by two straight stones set up on end, and sloping towards one another so as to form a primitive arch.

Norman.

We come next to the Norman period. In the first place it must be mentioned that great differences are to be found between the earlier and the later parts of this period ; differences which though apparent in detail rather than in style, are not the less worthy of notice. In the early part of the period the masonry joints are wide and clumsy, as is well seen in the greater part of the walls and the lower portion of the tower of Barton Seagrave; in the later part the masonry is finely jointed. Again, in the earlier years of the period, a lack of orna-mentation is to be noticed, whereas towards the close the work became more ornate, as in the beautiful tower of Castor. Mouldings were cut deeper than previously, pillars which had been square and solid became more ornamental (as at St Peter's, Northampton), and on the capitals foliage and figures began to be carved. The church at Maxey furnishes us with a good example of Norman work at

Saxon Chancel Arch: Wittering Church

different dates. The most characteristic feature of the whole period was the round head employed for windows, arches, and doorways, while the zigzag ornament began to be freely used. Almost the whole of the interior of Peterborough Cathedral belongs to this period. Norman work, we may add, is to be found in very many churches up and down the county, indicating a considerable population at that period.

Saxon Door in Tower of Barnack Church

Early English.

If we compare the churches of the next period, Early English, with those of the preceding one, we notice that they are lighter and more graceful. We find windows of the type known as "lancet-shaped," long and narrow; boldly projecting buttresses; roofs with an acute pitch; pointed arches, generally with rich mouldings;

and pillars with clustered shafts. Tracery is beginning
to be used, and the toothed ornament is frequently
employed. The east end of the church at this period was
almost invariably square, and the towers, which were
built to a greater height than in the Norman period,
generally ended in a "broach" spire, without any parapet
at its base, showing where the octagonal spire was joined
to the square tower. Examples of Early English work
are plentiful in Northamptonshire; in the south-west of
the county the towers of the churches at Brackley and
Moreton Pinkney belong to this period; the west door-
way and arcade of Canons Ashby and the nave arcade of
Towcester are also instances. Near Northampton we
may cite the church at Floore, which belongs mainly to
this period in structure; the tower, aisles, and nave
arcade of Great Brington; and the chancel of St Giles'
at Northampton. The church at Rothwell has a noble
transitional nave (Norman to Early English). Near
Thrapston are the following good examples:—Aldwinkle
All Saints', which is Early English in origin, and the
nave of Aldwinkle St Peter's; the tower of Raunds, and
the tower, south aisle, and porch of Stanwick; the churches
at Cotterstock and Higham Ferrers, which are mainly
Early English; the broach spire of Ringstead; the tower,
spire, and chancel of Polebrook; and the arcades of the
church at Oundle, the ground plan of which is almost
all Early English. In the north-east corner of the county
we may mention the southern arcade and porch,
belfry, and spire of Barnack, the west front of Peterborough Cathedral, and the church of Warmington,

Warmington Church; Nave with E. E. vaulted wooden roof

which still has its beautiful original roof, and is considered a magnificent example of Early English in all stages.
The church at Etton, we should add, is of particular interest, because, although it is simple in both design and
detail, it has been well described as "an absolutely
untouched specimen of good plain thirteenth century
work throughout."

Decorated.

The change to the next style, known as the Decorated,
was very gradual. The chief things to be noticed about
this period are the windows, which are large and divided
by mullions; the ornaments, which are numerous and
carved with great delicacy, serve some structural purpose; and the pillars no longer have detached shafts.
In the south-west of the county, the tower of Canons
Ashby and the tower and spire of Byfield afford us good
examples of Decorated work; near Northampton we
have the church of Easton Maudit, which is chiefly early
Decorated, the tower and nave of Great Billing, and the
nave of St Luke's, Wellingborough; in the neighbourhood of Thrapston we have the whole exterior of Corby,
the chancel, tower, and spire of Aldwinkle St Peter's,
windows at Oundle, the chancel of Cotterstock, and the
church at Finedon, which belongs wholly to the Decorated period with some peculiar details. In the north-east
of the county we may mention Ufford, in which are
some fine examples of Decorated work, both geometrical
and curvilinear, and the nave and aisles of Nassington.

Perpendicular.

One more period remains to be discussed, that known as the Perpendicular. The name gives us the clue to its chief characteristic, lines crossing one another at right angles. These can be seen on the surfaces of the building, and they are especially to be noticed in the tracery at the heads of windows, where horizontal and vertical lines take the place of flowing curves[1]. In the south-west of the county, we have the churches of Charwelton and Easton Neston which are Perpendicular throughout, and the tower and spire of King's Sutton. Near Northampton there is the church of Ashby St Ledgers, which is almost wholly Perpendicular; we have also the chancel and clerestory of Great Brington and the chancel of St Luke's, Wellingborough. Not far from Thrapston are the following examples:—the chantry chapel, tower, and sacristy of All Saints', Aldwinkle; the church at Islip, which is good uniform Perpendicular throughout; the particularly graceful tower and spire of Kettering, the tower of Tichmarsh, and the tower, spire, and nave of Brampton Ash. The tower, chancel, and transept arches of Oundle belong to the transition period between the Decorated and Perpendicular. In the north-east of the county the churches of Fotheringhay (with an octagonal lantern on the top of the square tower) and St Martin, Stamford, are wholly Perpendicular, and the eastern

---

[1] This was the period of careful design; the windows were made much larger, and we find clerestories sometimes which are almost a continuous range of glass.

Oundle Church

portion of Peterborough Cathedral belongs to this period. Whiston, near Northampton, is one of the latest examples (1534) of wholly Perpendicular work.

The late Mr R. P. Brereton, of Oundle School, pointed out three special characteristics with regard to the churches of north Northamptonshire: (1) the predominance of spires in the northern part of the county, a feature which distinguishes its churches from those in the southern part of Northamptonshire. In the northern part more than half of the churches have spires, and of the spires, two-thirds are of the "broach" kind. That at Warmington is famous for its richness of detail, that at Polebrook for its elegance of form; whilst in Aldwinkle St Peter's we have a beautiful example of graceful outline. Of those rising from parapets, the spires of Kettering, Deene, and Oundle are especially fine examples: (2) the frequency of Transition work, belonging particularly to the period called the "Geometrical," i.e. that from about A.D. 1260–1320. So many examples of this age of change of style—shown clearly in arches and doorways and especially in the growth of window tracery—exist within the county, that probably no other district could be found to illustrate better the change which was coming over church architecture in the years mentioned above : (3) the great interest and beauty of many whole churches, and of details in the large majority of them. Almost every church in the county possesses some point of interest or beauty, and will repay one for the time spent in visiting it.

Fotheringhay Church

17. Architecture—(*b*) Military.

From the earliest times men have chosen positions rendered impregnable, or at least partially secure, for the construction of forts or fort-like dwellings. The summit of a precipitous rock, an island in a lake or river, a piece of land within the loop of a river, are typical sites chosen for building a place of defence. In England, during the British and Roman days and throughout the Anglo-Saxon period, the chief defences of such a stronghold were the earthwork and the stockade. In Norman times a new structure appeared, the castle. At first this consisted merely of a " keep " with a walled enclosure attached to it for the protection of cattle; the keep being a tower whose ground plan was rectangular, with enormously thick walls. The walls were only pierced with very narrow holes for windows and the accommodation inside was most primitive. From this type of fort was evolved the castle, with its courtyards and embattled walls, round which, when possible, ran a moat, crossed in front of the principal gateway by a drawbridge. As military art progressed, considerable changes and developments took place, until the introduction of gunpowder made the defence of such a building nearly impossible. From that time onwards, and with more settled conditions of life, the fortified type of dwelling was used less and less ; and the manor house and country seat superseded the castle as a residence for wealthy men and their families.

Northamptonshire, it must be admitted, is not rich in old castles ; there were few points within the county of strategical importance and the consequence was that, with the exception of those at Northampton, Rockingham, and three places in the valley of the Nene, few military castles were erected within the county. In this chapter we propose to give a brief account of the five whose remains can be seen to-day. Two of these were built by the same man, and are therefore of nearly the same age ; we will consider them first.

The castle of Northampton was built by Simon de St Liz (or Senlis), said to have been the first Norman Earl of Northampton, probably during the reign of William Rufus. In a document of the year 1130 A.D. we find the record of a certain payment made by the King (Henry I) for land taken "into his castle" of Northampton. At some time therefore, before 1130 A.D., the castle must have passed into the royal possession. Its military importance can be judged from its position ; it was situated near the centre of England ; it lay near the road leading north from London by the West Pennine route ; and lastly it stood at the head of the valley of the Nene, commanding one of the important routes across England. On the west the castle was defended by the river; on the other three sides it had a moat 18 feet wide. This moat was crossed by a drawbridge which was guarded on its outer side by a barbican ; on the inner side stood the great gate of the castle, with its portcullis. Along the walls were situated towers to aid in the defence, and within rose the keep. Within its walls many interesting events took place, to which reference is made in our chapter on the History of the town. In 1248 and 1251 A.D.

we find it recorded that many improvements were made to the castle, including the addition of glass windows, an indication that at that time glazing was coming into use in such buildings. From documents of the time we learn that the castle contained two chapels, one for the King and the other for the Queen, and that in the later part of the thirteenth century a survey was made of the castle and park. Later, in 1321, commissioners who had been sent to examine into the general state of the castle recommended that extensive repairs should be undertaken, but we have no indication that these were carried out. From 1391 onwards the importance of the castle appears to have dwindled, and it seems to have been used chiefly as a county gaol. In the Civil War it was held on the side of the Parliament, but on the restoration of the Monarchy in 1660 it was "slighted," and its gates, walls, and fortifications were demolished. A disastrous fire occurred in the town in 1675, and the materials of the ruined castle were largely used for rebuilding the houses wrecked by the fire. Lastly, in 1876, the London and North Western Railway Company bought the land on which the castle originally stood ; and the carrying out of their plans meant the final destruction of what was left of the old castle. Nothing now remains but the postern gate (taken down and rebuilt by the Railway Company), and a piece of the old wall.

Simon de St Liz (or Senlis) built yet another castle—that at Fotheringhay. This passed through many hands before Edward III granted it to one of his sons, Edmund of Langley, Duke of York, who rebuilt it, constructing

the keep in the form of a fetterlock. Mary Queen of
Scots was brought to the castle in the month of Sep-
tember, 1586, and was kept a prisoner there until her
execution in the February of the following year. Fuller,
who lived in the seventeenth century, tells us that when
he visited the castle he found the following lines in Queen
Mary's handwriting scratched with a diamond upon a
window pane :—

> " From the top of all my trust,
>   Mishap hath laid me in the dust."

The castle was dismantled soon after a survey made in
the last year of James I, and the materials of which it
was composed were sold. Sir Robert Cotton bought the
great hall and removed it to Connington, and materials
from the ruins were used for various purposes in the
neighbourhood. Little is left to be seen now ; the great
mound on which the keep stood is still an object of
interest ; it is said to have been thrown up in pre-Norman
days to guard a ford crossing the river. The foundations
of the fetterlock tower remain, and a fragment of wall
lies between the mound and the river. It is believed
that originally there were two moats, but it is impossible
to describe the actual form of the castle as we have little
beyond tradition to guide us. Fotheringhay castle stood
in about the same position with regard to Ermine Street
as Northampton Castle did to Watling Street; it guarded
the left bank of the river and was built upon a spur of
ground jutting out into the valley, a position which
rendered it easily capable of defence.

Next in importance comes Rockingham Castle, built by order of the Conqueror on the ridge of high ground which runs along the north-east border of the county south of the Welland.   The castle was erected within earthworks thrown up long before, possibly by the British, and served a triple purpose—it guarded a spot where the

Fotheringhay Keep Mound

Welland could be crossed, it commanded traffic up and down the river valley, and it was a favourite resort for the kings, as it was situated in the midst of a great tract of forest which had always afforded excellent hunting. In 1094 the famous conference, attended by Anselm, took place within its walls.   We have records of visits paid to it by various kings, chiefly for the sake of the hunting, and of extensions and repairs which were carried out from time to time. One curious thing, pointed out by Leland in 1545, is worthy of record; the walls were embattled on both sides, so that if a storming party forced an entrance through either gate, the force holding the wall might still maintain the defence. Since Leland's

Barnwell Castle

time a great deal of the building has perished, but one of the great gates remains, flanked by two circular bastion towers pierced for the use of bowmen, and the keep mound is still to be seen. The gateway gives admittance to the outer bailey or court of the castle, in which are situated the buildings inhabited now. The whole area enclosed was about 3½ acres, and the surrounding wall (part of which still exists) was nearly 9 feet thick.

At Barnwell St Andrew's are the remains of a castle said to have been built by Berenger le Moyne about the year 1264. A few years later he was called upon by Edward I to show by what warrant he had built it; whereupon he hurriedly sold his right in the castle to the Abbot of Ramsey, who, with his successors, held it until the Dissolution. It passed, about 1540, into the hands of Edward Montagu, who is said to have put the building into a good state of repair; but it was not long before it began to fall into decay, and eventually it supplied building material for the neighbourhood.

It is difficult to attach any military importance to this castle, which was built up a small side valley opening on to the main valley of the Nene opposite Oundle. It consists of a quadrangular court with towers at the angles, entered by a gateway flanked by round towers.

Not far from Barnwell can be seen the remains of one more castle, that of Thorpe Waterville. The moats and foundations are still visible, but the materials of the original building have long since disappeared. It was built by one of the Watervilles, who held the land from the days of the Conqueror to those of Edward I. That it was at one time a well fortified place can be gathered from references made to it in documents of the fifteenth century.

18.   Architecture—(*c*) Domestic.

In the preceding chapter we traced briefly the development of the castle, and showed that the fortified house
consisted originally of a keep alone.   In this chapter we
have to deal with buildings whose first purpose is to afford
a home, and not merely a place of refuge.

We have pointed out before that the accommodation
of the keep was of the most primitive character.
Rectangular in shape, the keep was several storeys high,
and its walls were very thick.   As a rule there was
only one room to a floor, and this was draughty and
badly lit by reason of the narrow unglazed wall-openings
which served as windows.

The first suggestion of a dwelling-house, as distinct
from a castle, is seen in the fortified manor house.
The chief room in this was the hall—indeed we may
say that the hall was the house ; even to-day we still
often find the principal dwelling in a parish called
"The Hall."   At its upper end was a raised platform,
or daïs, on which the lord and his family took their
meals, and in the one great room the whole household
lived and slept.   As time went on, it became usual to
have a smaller room, known as the "Solar" at the daïs
end of the hall for the lord, and this developed slowly
into a suite of rooms for the family ; at the other end
was situated the kitchen, which gradually expanded into
a set of domestic offices and quarters for servants.   The
suite of rooms for the family and the servants' quarters

by and by developed into the wings of the building, with the hall in the centre connecting them. The fireplace in the main building was frequently in the middle of the floor; we have instances of this as late as the sixteenth century, for at Deene Hall, which was built in the reign of Edward VI, there was no fireplace with a flue for the smoke until one was constructed in the nineteenth century; moreover the opening through which the smoke originally escaped can still be detected. The keep had not disappeared altogether; at Longthorpe, near Peterborough, can be seen the remains of one built in the days of Edward I, joined to and forming an integral part of the house. The purpose of defence was secured by a moat, which was crossed by a drawbridge leading across to a strongly-defended gateway. Woodcroft Castle, near Helpston, is a moated house belonging to this period of development. It consists of two wings, one much longer than the other, meeting at right angles, with a tower in the corner so formed. Both wings are probably imperfect; the shorter is of later date than the longer and there was once a circular tower at the end of the latter. Glazing, of course, was unknown at first, but from orders issued for the insertion of glass windows in royal houses, we can tell that it was becoming the practice to use glass towards the close of the thirteenth century.

With the advent of the fourteenth century, a general advance is noticeable in the comfort of dwellings; the need for defence was slowly passing away, and in proportion as this grew less, so comfort increased. The hall still remained the chief apartment, but the smaller

rooms were beginning to assume more and more importance. The design of the whole building was still very
haphazard, and the idea of aiming at symmetrical arrange-
ment was only in the birth. Part of the house at Drayton
was built by Simon de Drayton in 1328, but the original
open woodwork roof of the hall is now hidden by a
ceiling of later date. A smaller building of the same time
is the manor house at Northborough in the extreme north
of the county; in this the old hall remains, but it is
divided now into two storeys.

In the buildings of the fifteenth century a greater
effort can be discerned on the part of the designers to
produce architectural effect; they still paid a good deal
of attention to the need for defence, but the accommodation for the inmates increased considerably. The tendency to build symmetrically also became more marked.
An example of this period can be seen at Fawsley, the
hall of which contains one of the fine bay windows which
were beginning to become general in this century.

In the sixteenth century (and especially during the
second half of it) a remarkable development in building
took place, which may be ascribed to the fact that it was
a period of wealth and peace. The character of the work
was strongly affected by the movement known as the
Renaissance. The four most noticeable features of the
new architecture were the following :—ornament, which
had been rare in the past, became general ; the pointed
arch gave way to the semicircular ; windows became
square-headed ; and symmetry became more frequent in
the plans of houses. Notwithstanding these changes, the

Northborough Manor House

body of the house still adhered to the old English type.
Houses came to be built more as homes than as places
of defence; they were still generally surrounded by a
moat, but this was crossed now by a permanent bridge.
A characteristic of Tudor houses was a tower over
the entrance. The hall was still the most important
apartment, but its day was passing, for the dining-parlour

Kirby Hall; interior of Courtyard

had been introduced amongst the family suites of rooms.
By the end of the first quarter of the seventeenth century
the importance of the hall, as the chief living room, had
ceased.

Northamptonshire is rich in examples of Elizabethan
houses. The great house at Burghley was built partly in
Mary's reign, partly in the reign of her successor. When

the estate passed into the hands of Sir William Cecil (afterwards Lord Burghley) in 1553, there was a house already in existence, but he at once set about effecting great changes and additions. The great hall and the kitchen were the first buildings due to him, and between the years 1577 and 1587 he completed the rest of the courtyard. Kirby Hall is another fine Elizabethan house,

Lilford Hall

built round a great courtyard, and remarkable for the beautiful detail in the stone-work, particularly in the pilasters and cornices. It is now a hopeless ruin, but weathering has hardly affected the sharp outlines of the carving on the stone. The largest of the Elizabethan houses in this county was that at Holdenby, built by Sir Christopher Hatton, Elizabeth's Lord Chancellor, between the years

1575 and 1585. It was in the form of two quadrangles and measured 360 feet by 224 feet. A small portion of the main building, which was in a state of ruin for a long time, was restored in 1875 and 1888, but the two gateways which stood in front of the house still

Lyveden "New Build"

remain. Castle Ashby, begun in the reign of Elizabeth and possessing additions by Inigo Jones, parts of Rushton Hall and of the house at Deene Park, and a wing of Drayton are also Elizabethan buildings of interest. One of the most charming houses of the sixteenth century in the whole county is Lilford, beautifully placed on the

Nene, and noted for its fine stone work. Many unpretending manor houses such as Lyveden "Old Build," and Pilton Manor House (now the Rectory) are of about the same age.

In the seventeenth century the pursuit of the Italian ideal became still more marked ; it was the age of Inigo Jones and Sir Christopher Wren, who left a permanent mark on English architecture. The two most distinctive characteristics of the buildings of their time are the absence of gables, and the substitution of sash windows for the old mullioned ones. A typical plan of the day was to place all the principal rooms of a house in one lofty central block, and to flank this on each side with the less important rooms ; the latter, however, were placed some distance away, and were joined to the central block by colonnades. Such a grouping was carried out by Inigo Jones at Stoke Bruerne, which was built between 1630 and 1636 ; the central block, unhappily, was burnt down in 1886. Cottesbrooke Hall, although of later date, is a good example of this type of plan. Thorpe Hall, near Peterborough, is a fine example of the middle of the seventeenth century ; it was built during the Commonwealth, and was designed by John Webb, a pupil of Inigo Jones, whose influence is very obvious.

The eighteenth century was an age of stateliness within and without, and comfort was ruthlessly sacrificed in the effort to produce fine effects. Boughton House was built in this period, but it is hardly a good type of the architecture of that day. The original house had been built in the middle of the sixteenth century, and the

Drayton House

Duke of Montagu, after his return from France at the close of the seventeenth century, rebuilt his home in imitation of the palace at Versailles. It has suites of rooms which open one into another, without a passage. Another house with much work of the period is that at Drayton. We have already referred to it as dating from the

Old Gates—formerly at Oundle, now removed to East Haddon

fourteenth century, but a great deal of the building is work of the eighteenth. A new front was given to the hall, whose open timber roof received a ceiling at the same time ; sashes were substituted for the old mullioned windows, and cupolas on columns were placed upon two old towers. The house at Althorp also belongs to the eighteenth century, but it has been recased in modern times. One good feature of the period is the excellence of the ironwork—the railings and gates which are to be found at the boundaries of gardens. Drayton furnishes beautiful examples of these.

The details peculiar to the different buildings which we have described are very similar to those in churches of the corresponding periods. In the earlier years, however, there was far greater simplicity in domestic than in ecclesiastical building, because houses of the former type were liable to attack.

There were four ways of treating the interior surfaces of the walls in these early houses; the first was to leave them bare; the second was to plaster them, but so thinly that, although crevices were closed, irregularities were not hidden; the third method was to cover them with wainscot, i.e. with oak panelling; the fourth method was to hang them with tapestry. In Elizabethan and Jacobean houses this panelling was considerably developed, and the plastering became much more ornamental; as examples in this county we may cite the drawing-room ceiling at Canons Ashby, several very fine ceilings at Apethorpe Hall, and one or two at Deene. Magnificent chimney-pieces, sometimes of stone, sometimes of wood, and of very varied design, are often found in these houses. In the eighteenth century all these interior features became still more elaborate.

The type of building usual in any district is, of course, largely influenced by the material at hand. No one passing through Northamptonshire can fail to be struck with the prevalence of stone as a building material; this is not to be wondered at when we remember the great bed of Oolite which stretches through the county. The stone is easily worked, and weathering gives it a soft grey hue which is very pleasing to the eye. All the older houses, and many modern ones, are roofed with Colly Weston slates, of which a plentiful supply has always been forthcoming. The roofs of old stone cottages, bent

Cottages with typical chimneys

with the weight of years as well as with the burden of their heavy stone coverings, are a characteristic feature of the villages of the shire. Many of these humbler houses, moreover, exhibit points of architectural interest—mullioned windows, good simple chimneys, and quaint gables. The stone is pressed also into the service of garden and boundary walls, which may very commonly be seen built with stones compacted together with mud, or with mortar little better than mud, and capped with tiles or cement to prevent the rain soaking into the wall and causing it to swell and fall.

19. Communications—Roads, Railways, Canals.

We have already on several occasions referred to some of the main roads of Northamptonshire; we must now examine them more closely.

Amongst those made before Roman days is the Welsh Road, or Banbury Lane, which is certainly one of the oldest, if not the oldest highway in the county. This road probably began at Clifford's Hill, an early earthwork on the south side of the river Nene about three miles east of Northampton, and ran westward to Hunsbury Hill. This was a British camp on the brow of the hill about two miles south-west of Northampton, and from a military point of view no doubt always a most important position. From this camp the road follows in the main a straight line leading in a south-westerly direction to Banbury, where it leaves this county and goes through Tadmarton Camp to the remarkable Neolithic stone circle known as Rollright Stones. For a good part of its course this road forms the boundaries of the parishes by which it passes. The track was during the Middle Ages much used by drovers for bringing cattle from Wales to the Midlands, as the great width of the road (from 40 to 80 feet) afforded ample pasturage for the beasts, and in later times the absence of toll-gates was a matter of consideration.

Another old road is that leading from Northampton north-westwards to Welford, which has many of the characteristics of Banbury Lane. Where it crosses valleys, as between Kingsthorp and Brampton, and between Brampton and Spratton, it winds round the slope of the hill until it reaches a narrow point of the valley, then turns sharply, and after crossing the river at the narrowest point, turns again and pursues the original course. When it once reaches the crest of the range of hills near Spratton, it keeps on the summit until it reaches Welford.

These are the two chief roads of those made in pre-Roman days. We will now turn to those made in Roman times and later.

The two main roads which traverse Northamptonshire from south-east to north-west are the London to Chester road which crosses the county in its broadest part, and the Great North Road which crosses it where it is narrowest. The road from London to Chester coincides with the Roman Watling Street from Stony Stratford to Towcester and thence to Weedon, where it leaves the old Roman road and branches away to the west, passing through Daventry and leaving the county near Braunston. From Weedon, Watling Street holds on in practically the same straight line, but beyond Watford it is now grassed over for more than two miles. The Great North Road enters Northamptonshire at Wansford, and crossing the end of the ridge which runs along the

Watling Street
(*Part near Crick, now unused*)

north-west border of the county, skirts Burghley Park and drops to the valley of the Welland at Stamford. Ermine Street, the great Roman road from London to Lincoln and York, is followed continuously by the Great North Road for a course of 56 miles to a point a little north of Chesterton in Huntingdonshire, but at that point the two

On the Great North Road

tracks separate. The Great North Road turns to the west; Ermine Street ran straight on, passing over the site of an old Roman town near the modern Castor. It crossed the Nene into what is now Northamptonshire by a bridge resting on stone piers which were removed when the river was made navigable. From this point its course can be traced partly by the elevated ridge on which it ran, partly by the remains of its ancient foundations, past Southorpe, Walcot Hall, and Burghley Park, where it descended to the valley of the Welland, crossing the river half a mile above Stamford.

Thus of the two great Roman roads which crossed our county, one is still used in part, the other has perished altogether. Each of them is remarkable for the directness of its course—no turning, no flinching from the hills. The Roman road was an epitome of the character of the Roman soldier.

Of post-Roman roads an important one is that which leaves Watling Street near Stony Stratford, and leads north to Northampton. The London to Nottingham road, through Bedford, enters our county near Rushden and passes through Higham Ferrers and Kettering. Another road, that from Huntingdon to Market Harboro' and Leicester, passes through Thrapston and Kettering. From Northampton two important roads lead out in a northeast direction; one runs down the valley of the Nene, connecting the county town with Peterborough; the other goes, roughly speaking, along the north-west ridge, passing through Kettering, Weldon, and Colly Weston, connecting Northampton with Stamford.

The waterways of the county, rivers and canals, must claim our attention next. The two main rivers are, of course, the Nene and the Welland. The latter is not of such importance to the county as the former is, or might be; it only runs along the border of Northamptonshire and so cannot rival the Nene in utility; moreover it is only navigable for small craft up to Market Deeping.

The Nene has always been a great waterway, but in early times when roads were bad and practically impassable in bad weather, it was of far greater importance than it is to-day. Its valley, however, has always been subject to serious floods. In 1848 there occurred a flood which covered from 8000 to 10,000 acres of pasturage between Northampton and Peterborough. A committee was formed to consider what means ought to be taken to prevent the recurrence of this grievance. The final outcome was the appointment of the body known as "the Nene Commissioners," which is composed of all those estate owners (or their representatives) who possess more than 50 acres of land subject to flooding by the Nene or its tributaries. The money necessary for keeping the main river navigable and for draining the land on each side is raised by taxing the land which is subject to being flooded. The original scheme was to deepen the channel from Wisbech to Peterborough, but the engineering difficulties proved so great that after many thousands of pounds had been spent, the scheme had to be abandoned.

The basin of the Nene was then divided into three districts—(1) the region above Northampton where navigation is impracticable ; (2) the region between Northampton bridge and Peterborough bridge ; (3) that from Peterborough bridge to the sea. The duties of the Commissioners are twofold :—firstly to keep the river open to navigation (except in the first district mentioned above), and secondly to take measures to ensure the efficient drainage of the land bordering the river.

With respect to navigation, the district from Peter-borough to Sutton Docks presents many difficulties; that between Peterborough and Northampton is said to be better than it has ever been before. The principal work has been the construction and upkeep of locks and staunches, and the removal of shoals. The Commissioners some fifteen years ago had a steam dredger built of such a design that it can pass through the locks of the second region, and they spend a sum of between £200 and £250 annually in dredging alone.

To promote good drainage, the chief care of the Com-missioners is to keep the main channel of the river open, that it may be able to cope with the water when the gates are open in times of heavy floods. For instance, in summer it is important to prevent the river from becoming choked with weeds. Straightening the course of the river in many cases would be a great help, but there are diffi-culties in the way, one of the chief ones being that the river in so many places forms a boundary, sometimes between estates, at others between parishes or counties. The Com-missioners control all tributary streams, and to encourage the owners to clean out the water-courses on their land they often contribute a part of the expense incurred.

Turning now to canals, we find that the Oxford Canal passes through a small portion of the county near Braun-ston, from which place the Grand Junction Canal runs eastward under a tunnel to the north of Daventry, near which are two large reservoirs which supply the canal with water. Near Norton a branch runs north to Foxton, where it joins the Leicestershire and Northamptonshire canal, thus forming a connection with the Trent and Mersey systems. After passing Norton the Grand Junction Canal turns south-east and runs to Blisworth, from which town it sends a branch to Northampton, joining the Nene and permitting communication with the North Sea. From Blisworth it runs south through a tunnel and proceeds to Buckingham. Thus from

Grand Junction Canal Tunnel, through hills near Daventry

Northampton communication by water is possible (1) with the North Sea by the Nene; (2) with the Severn by the Oxford Canal; (3) with the Birmingham system by the Oxford Canal; (4) with the Trent by the Grand Junction and Union Canals; (5) with the Mersey by the Grand Junction Canal; (6) with London by the Grand Junction Canal.

The chief argument in favour of water-traffic is its cheapness, the most serious drawback is its slowness; in illustration of the latter it may be pointed out that it would take a barge three days to travel the 61 miles from Northampton to Peterborough by the Nene, whereas the railway can carry goods from one town to the other in two hours. Another objection in regard to river traffic (especially important in a case like that of the Nene) is the danger of having all navigation stopped by floods. A Royal Commission has recently been considering the whole question of canals, which have, to a great extent, fallen into disuse through the competition of the railways. From the final report of the Commissioners there is no doubt, however, that a development of the Nene waterways would be advantageous in many ways, "not only because it would afford an exit into the general system of English waterways for the penned-in navigators of the Fens, but because round Peterborough (chiefly in Huntingdonshire) there is a vast brick-making industry, which, if there were cheap transport via Northampton and the Grand Junction Canal, or to the port at Wisbech, and so by sea to London and other places, could contribute a limitless amount of cargo, of a kind for which water transport is eminently fitted."

Turning to the railways, we must bear in mind the fact that Northamptonshire lies across the direction of the great routes from London northwards. Four of these traverse our county; the Great Central passes through the south-west corner; across the west runs the London and North Western, which passes to the west of the

Pennines; through the centre runs the Midland, which climbs the Pennines by the valley of the Aire; across the east runs the Great Northern, which chooses the "East Coast route." If to these we add the branch of the London and North Western which runs down the valley of the Nene, connecting the last three of the great routes to the north mentioned above, we have the railway system of Northamptonshire—four lines running across the county and one line running down it.

The London and North Western main line enters the county near Ashton, and traverses it in a north-west direction, passing through Blisworth and Weedon on its way to Rugby. The line runs through a tunnel over a quarter of a mile long under a spur of the limestone hills near Stowe (and at the same time under Watling Street), and under the main range of hills by Kilsby tunnel, $1\frac{3}{4}$ miles long. Between Roade and Blisworth, a branch line leaves the main track and runs through Northampton, rejoining the main line at Rugby. The south-west of the county is served by the line from Northampton to Banbury, which passes through Blisworth and Towcester. We may add that Northampton is joined by direct railway routes to Bedford and Leicester.

The Midland main line enters Northamptonshire near Irchester, crosses the valley of the Nene at Wellingborough, and passes up the valley of the Ise to Kettering. It crosses the main ridge of hills at Desborough and descends to the valley of the Welland at Market Harborough. Two of its branch lines deserve notice; one comes from Huntingdon through Raunds and Thrapston, joining the

main line south of Kettering; the other breaks off south
of Rushton, and passing through Geddington, Weldon,
and Gretton, crosses the valley of the Welland at Harringworth by a remarkable brick viaduct 1290 yards long.
It is formed by 82 arches (of 40 feet span), and the level
of the rails is 64 feet above the ordinary water level of
the Welland.

Viaduct across the Welland Valley

The Great Northern main line crosses the county at
its north-east end where it is very narrow, and passes
through Peterborough, whence branch lines run to Stamford and Spalding.

From Northampton down the valley of the Nene runs
a branch of the London and North Western railway which

links together all the towns which lie in the principal valley of the county. The line is built upon an embankment for most of the way to render it safe from floods, and the winding of the river makes it necessary for the railway to cross it again and again. Two branches from Wansford connect the valleys of the Nene and the Welland, one by Barnack and Stamford, the other by King's Cliffe. At Peterborough a junction is effected with the Great Eastern which has a line from there to Ely and Cambridge.

20. Administration and Divisions.

In pre-Norman days, the smallest unit in the government of the country was the township, the principal man of which was called the "reeve." Every freeman had the right to attend the assembly of his township, which settled small matters, but passed on the more important cases to the Hundred-moot (or Hundred-court). The "Hundred" in all probability originally comprised a district inhabited by the families of a hundred warriors. At the head of this division was the "Hundred-man," by whom the Hundred-moot was summoned. This met once a month and was made up of four men and the reeve from every township, and the Eorls and Thegns living in the Hundred; it settled questions of property and decided criminal cases. The Shire included several Hundreds. Its chief officers were the Ealdorman (appointed by the Witan), and the Shire-reeve (Sheriff), so called to distinguish him from the humbler reeve of the township. The duties of the latter were to attend to the rights of the king in the Shire, and to see that the decisions of the Shire-moot were carried out. This Shire-moot, or Shire Council, was evolved from the Folk-moot, the purpose of which had been to administer justice between people living in different Hundreds; it was attended by the freemen and Thegns of the Shire, and settled all the more serious matters affecting its own county. In the Witan, or Council of the Nation, the Shire was represented by the Ealdorman and the Bishop. The latter attended on behalf of the church, which, it must be remembered, exercised a great influence in those days on the general affairs of the country.

The history of the Hundreds of Northamptonshire has been carefully traced out; we have lists of the Hundreds, and particulars of the land in each, in several interesting documents, one compiled in the Conqueror's reign before the year 1076, a second in Domesday Book in 1086, a third in the twelfth century, and a fourth at the beginning of the fourteenth century. Changes occur in these lists for various reasons, one obvious one being that part of what is now Rutland was included in Northamptonshire in the eleventh century. The number of Hundreds (counting Nassaburgh or Peterborough Liberty as one) at the present time is twenty.

When the Normans came to England, they found the " manorial " system virtually established. The "manor" was practically the old township, which had passed, under a species of feudalism, into the hands of a lord. The town-reeve had been displaced by the lord's steward, and the manorial courts had taken the place of the town-moots.

The institution of Royal Courts of Law, of Judges who went on circuit, of trial by jury, are matters of history which we need not deal with here. Many developments took place as the years went by, but it is to be noticed that the general tendency, under the feudal system, was to centralise all government, that is to say to gather the reins of power more and more into the hands of the king.

We must turn now to the modern methods of administration in use in Northamptonshire. The county is divided into districts known as Poor Law Unions, each of which takes its name from the most important town which it contains; the whole of Northamptonshire is included within the area covered by 16 unions, but four of these lie almost wholly outside the county, while some of the remaining 12 include parishes situated in neighbouring counties; in each of these districts a body of men are elected who form what is known as a "Board of Guardians." Their chief duties are to appoint "relieving officers," i.e. officials to administer relief under the Poor Law, and to manage the Workhouse in their Union. The smallest unit, corresponding to the old township, is the Parish, but its Council has very limited powers, and most of the necessary business of each parish is transacted by the Rural District Council in whose district it happens to be. This latter Council, which has authority over all parishes within its district except those under the control of Urban District Councils, is responsible for the upkeep of the roads (other than the main ones), the proper regulation of sanitation works, water-supply, and other matters of public health. An Urban District Council has control of a parish upon which Urban powers have been conferred by the Local Government Board, and is responsible for such matters as the water-supply, the upkeep of roads within its area, drainage and sanitation, the lighting of streets, scavenging, etc. These three Councils, which we have mentioned, were established by Act of Parliament in 1894, and the important duties of the last two are carried out by means of money raised by rates, which they are empowered to levy upon occupiers in their districts.

We come next to the central administrative body, the County Council, which corresponds to the old Shire-moot. In the first place we must point out that for all purposes except Parliamentary representation, the Soke of Peterborough forms a separate county. The borough of Northampton also, for administrative purposes, forms a county by itself, and its municipal corporation has the powers of a County Council. We have then, within the borders of Northamptonshire, two distinct County Councils, one of which administers the affairs of the Soke or Liberty of Peterborough, and the other those of the remainder of the county excluding the borough of Northampton, to the powers of whose municipal corporation we have already referred. It will be sufficient for our purpose to describe the constitution and duties of one of the bodies—that which controls the affairs of the county area excluding the Soke of Peterborough and the borough of Northampton. It is composed of a chairman, aldermen, and councillors, the last named representing the various electoral divisions into which the county is divided. Its chief duties are in connection with the upkeep

of the main roads and bridges, education, matters of
public health, county buildings, asylums, diseases of

The Town Hall, Northampton

animals, the acquiring of land for small holdings, old-
age pensions, and the police. It appoints coroners for
the various districts, and maintains a county Medical

Officer of Health and other officials to assist it in its manifold duties. This system of central government, we may add, was established by Act of Parliament in 1888.

With regard to the upkeep of the main roads of the area for which the County Council is responsible, the practice has hitherto been to allow each District Council to keep in repair the main roads within its district, the expense being borne by the County Council, but now the Council are gradually taking over the upkeep of the main roads.

For the carrying out of the Old Age Pensions Act of 1908, a "Local Pension Committee" has been formed, which has divided the county into twenty-one districts, for each of which it has appointed a sub-committee. To these sub-committees it has delegated its powers as much as possible.

The educational work of the Council is carried out by a special committee consisting of thirty members, twenty-four of whom are members of the County Council, the remaining six being selected from outside. One of these six is appointed to represent University Education, another to represent Secondary Education, and two to represent Elementary Education; the other two places are allotted to women. This Education Committee possesses the full powers of the County Council with respect to education, "except (1) the power to raise a rate or borrow money; (2) the appointment of the secretary for Education; (3) the creation of any new administrative office; (4) the approval of any settlement as to the application of any Endowment or Endowments."

Under the Education Act of 1902, the right was reserved to towns of a certain size to control their own Elementary Education; Kettering exercises this right, which belongs of course to the borough of Northampton also. With these exceptions, the whole of the Elementary Education of the county is controlled by the County Council, who also make provision for Secondary Education and maintain the Evening Schools.

For the administration of Justice the county is divided into nine Petty Sessional Divisions, each of which has its own Court of Petty Sessions, presided over by the Justices of the Peace resident in the division. These courts have powers of "summary jurisdiction" in such cases as those of assault, minor cases of theft, offences against the Game Acts, etc.; but more serious matters are passed on, after a preliminary investigation, to the Courts of Quarter Sessions, or else to the Assizes. Courts of Quarter Sessions are held four times a year at Northampton; their business is mainly criminal—they were in fact established to relieve pressure on the Assizes. Cases of murder are sent on to the Assize Courts, but it is an interesting fact that the chairman of Quarter Sessions at Peterborough (where they are held as well as at Northampton) still possesses the right of passing the sentence of death in a murder case. This right, we may add, has not been exercised since 1812. The Assize Courts are held at Northampton three times a year; a Judge of the High Court always attends these, and his jurisdiction in all civil and criminal cases is unlimited.

We have referred above to some of the duties of a

Sheriff in bygone days; the office carried with it many
other duties, but most of these have now lapsed.　The
Sheriff still attends judges on circuit, and at parliamen-

The Guildhall, Peterborough

tary elections he acts as returning officer.　His place as
military leader of the county was taken, in the reign of
Henry VIII, by the Lord Lieutenant, whose chief

function now is to represent the king in the Shire. Under the present military system, the Lord Lieutenant takes an active part in the organisation of the Territorial forces within his county.

Northamptonshire returns seven members to Parliament; two sit for the borough of Northampton, and one for the city of Peterborough; the remaining four are county members, and sit respectively for divisions named North, South, East, and Mid.

The whole of the county, with the exception of the parish of Thurning, is situated within the diocese of Peterborough, and is divided into two archdeaconries, those of Northampton and Oakham.

It is interesting to notice the tendencies that have been at work in the changes which have taken place in county administration. In the earliest times, each district managed its own affairs, but under the feudal system government tended to become centralised, firstly in the hands of feudal lords, and through them in the hands of the king. In more modern times another influence has been at work; the task of administration has passed back to the shires, but it has been centralised there in the county towns, whose work is carefully watched over by the central government offices in London.

21. The Roll of Honour of the County.

We shall try in this chapter to give some account of the people born in Northamptonshire, or closely connected with the county, who have become famous. It would of course be impossible to mention every one with some title to fame; we must content ourselves with the better known names.

In the year 1362 Henry Chichele the son of a yeoman, Thomas Chichele, was born at Higham Ferrers. When quite young he is said to have attracted the attention of William of Wykeham, who caused him to be sent to his college at Winchester, and afterwards to Oxford. After leaving the University he held several livings, and in 1405 he was sent on a mission to Pope Innocent VII with Sir John Cheyne—the first of several diplomatic missions in which the king employed him. In 1408 he was consecrated Bishop of St David's, and six years afterwards was promoted to the See of Canterbury. He contended against Lollardism and was a consistent supporter of a national church. He revisited his native town of Higham Ferrers in 1424, when he was holding a visitation of the diocese of Lincoln, and he then dedicated and endowed a college in the town where he had been born. He also built and endowed a hospital for twelve poor men. His old University owes to him the foundation of All Souls'. He died in 1443.

Throughout most of the fifteenth and sixteenth centuries the manor of Rushton was in the hands of the Treshams. Four of these Treshams deserve notice, three of them bearing the same name, Thomas. The first Sir Thomas acquired the manor about the year 1428; he was a devoted adherent of Henry VI, and suffered much for the royal cause. In the parliament of 1459 he sat for Northamptonshire, and was elected

Speaker; he represented the county again in 1467, and fought on Margaret's side in the battle of Tewkesbury 1471, where he was captured and executed. His grandson, also Sir Thomas Tresham, was knighted some time before 1530, and was Sheriff of Northamptonshire three times. He it was who in 1540 acquired the land round Lyveden. He was a staunch supporter of Queen Mary, who appointed him Grand Prior of the Order of St John in the year 1557. He died in 1559. His grandson, the third Sir Thomas Tresham, is probably the best known of the family. He was born in 1543, and when only fifteen years old inherited the Rushton and Lyveden estates. He was brought up a Protestant, but became a Roman Catholic and endured much persecution because of his belief. He was often thrown into prison, but in the intervals he erected buildings with which his name will always be associated—the market-house at Rothwell, the triangular lodge at Rushton, and the "New Build" at Lyveden. He died two years after the accession of James I. His son Francis was concerned in the Gunpowder Plot, and it appears certain that it was he who wrote to Lord Monteagle warning him not to enter the Houses of Parliament. He was arrested and, had he not died in prison, would have been executed.

With Grafton Regis is associated the name of the Wydevilles (or Woodvilles), and the name can be traced in the neighbourhood as far back as the twelfth century. Sir Ralph Wydeville, afterwards Earl Rivers, married the Duchess of Bedford; and Elizabeth his eldest child, who was born about the year 1437, became the wife of

Triangular Lodge, Rushton

Edward IV. One of her brothers was Anthony Woodville, Baron Scales and second Earl Rivers; he was a most
accomplished knight, much in favour with the king, his
brother-in-law, but on the death of the latter he fell a
victim to Richard III in 1483.

The manor of Ashby St Ledgers passed into the
hands of the Catesbys in the reign of Richard II. One
of the family, Sir William Catesby, possessed great influence with Richard III, who made him Chancellor of
the Exchequer. He it was who was alluded to as "the
cat" in the well-known doggerel lines of his day :—

> "The rat, the cat, and Lovell the dog,
>   Rule all England under the hog."

He was taken prisoner at Bosworth Field in 1485 and
executed three days afterwards at Leicester. The best
known of these Catesbys, however, was probably the
Robert Catesby who was one of the ringleaders in the
Gunpowder Plot in the year 1605.

At Fotheringhay Castle in 1452 was born a future
king of England, Richard III. After the battle of
St Albans in 1461 he was sent abroad for safety, but
was brought back and created Duke of Gloucester
when his brother Edward IV had secured the throne.
How he made his way to the throne, only to lose it and
his life after two troublous years, is known to all.

Richard Empson was the son of Peter Empson of
Towcester. He received an education for the bar and
soon became eminent as a lawyer. In 1476 he bought
estates in Northamptonshire, and in 1491 represented the

county in parliament. During the reign of Henry VII he was employed with Dudley by the king in most unpopular work—that of collecting taxes and penalties for offences against the Crown. This employment was bound to make him many enemies, and on the king's death he was committed to the Tower. He was tried and convicted at Northampton in the year 1509 and executed on Tower Hill in the following year.

In 1544 the Lord Mayor of London was Sir William Laxton, a prominent member of the Grocers' Company. He was born at Oundle, and, as Fuller tells us, was "bred a grocer in London." He died in 1556, having left his mark in Oundle by founding an almshouse and a school, both of which are still maintained by the Grocers' Company.

The best known name in connection with Northamptonshire is that of William Cecil, Lord Burghley. He was the only son of Richard Cecil, and was born in 1520 at Burghley. He was sent for his education to the Grammar schools of Stamford and Grantham, and to St John's College, Cambridge, which numbered Roger Ascham amongst its fellows at that time. From Cambridge his father sent him to Gray's Inn, where he applied himself to legal study. His political career began in the reign of Edward VI when he came under the notice of the Protector, Somerset. He represented Stamford in the parliament of 1547 and acted as secretary to the Protector, but was thrown into prison when the latter fell. He was too able a man, however, to be dispensed with, and in 1550 he was appointed one of the Secretaries of

State and made a Privy Councillor. Knighted in 1551 he succeeded in the following year to his father's estates and became master of Burghley. In Queen Mary's reign he was in retirement, but he was now a very wealthy man and spent huge sums on his houses at Wimbledon and Burghley. When Elizabeth succeeded to the throne she at once chose him as her chief Secretary of State. From that date onwards to his death he served his country with extraordinary sagacity. He was made Baron of Burghley in 1571 and died in 1598.

Another well-known figure in the days of Elizabeth was Sir Christopher Hatton. He was the second son of William Hatton of Holdenby. Born in 1540, he was sent to Oxford at the age of fifteen, and four years later was admitted to the Society of the Inner Temple. His appearance stood him in good stead, for the Queen took a fancy to the tall and handsome young man, made him one of her gentlemen-pensioners, and showered many favours upon him. He was the representative of Higham Ferrers in the parliament of 1571 and again in 1584, and was a member of the Commission which sat to judge Mary Queen of Scots at Fotheringhay. His connections with the county were manifold; he owned Holdenby and he bought Kirby Hall, though he does not appear to have seen much of the latter; he received from the Queen the manor of Parva Weldon and the demesne of Naseby; and he was also made Keeper of the Forest of Rockingham. Elizabeth made him Lord Chancellor in 1587, but he possessed little or no legal knowledge to fit him for the post. He died in 1591.

Robert Browne, the father of modern Congrega-
tionalism, was born at Tolethorpe Hall, a mile from the
border in Rutlandshire, and was rector of Thorpe
Achurch from 1590 to 1633. He was a well-known
Puritan, but it has been proved that he retained his
nonconformist principles while rector of Achurch. He
established and ministered to a separate congregation " of
the more forward spirits " in a building which still stands
in the village of Thorpe. After repeated persecutions for
nonconformity he died in Northampton in 1633.

Thomas Fuller was the son of the rector of St Peter's,
Aldwinkle. He was born at the rectory in 1608 and
received his early education at first from a school in the
village, afterwards from his father. At the age of thirteen
he entered Queens' College, Cambridge, taking his degree
in 1624. He held various cures, but is best known by
his writings; he is said indeed to have been the first
author who succeeded in making an income by his pen.
His first book was published in 1631 and was entitled
*David's Hainous Sinne, Heartie Repentance, Heavie Punish-
ment*. In 1642 he published his *Holy and Profane State*,
and in 1655 his *Church History*, which had taken him
many years to prepare. He died in 1661 before the
publication of *The Worthies*, but this task was carried
out by his son John Fuller.

James Harrington, visionary and republican, who
nevertheless attended Charles I on the scaffold, was born
at Upton in 1611, and is now chiefly remembered as the
author of *Oceana*, a romance in defence of republicanism.

In the rectory of the neighbouring parish of All

Saints', Aldwinkle, John Dryden the poet was born in
1631.   He was a scholar of Westminster and afterwards
of Trinity College, Cambridge.   In the year that he

John Dryden

graduated (1654) he lost his father and succeeded to a
small estate in Blakesley.   Three years later he took up
his residence in London, where he soon obtained literary
employment.   His work *Astraea Redux* greeted the

Restoration, but it is by his satires *Absalom and Achitophel*, *The Hind and the Panther*, etc., that he is best known. For some years he devoted himself to the drama, but the work was not congenial and was only undertaken for the sake of making money. He became Poet Laureate in 1670. In the last ten years of his life he produced his famous translation of Vergil and the *Fables*. He died in 1700 and was buried in Westminster Abbey.

John Bridges was born at Barton Seagrave in 1666. He took up legal work and was made solicitor to the Customs in 1695, subsequently holding other government appointments. It was in 1719, the year after that in which he became a Fellow of the Society of Antiquaries, that he began to collect materials for his great work on Northamptonshire. He himself travelled through every part of the county, and in addition employed people to collect information, to make drawings, and to copy records. In this way he spent a very considerable sum of money. It is said that he intended once more surveying the county himself, but he died in 1723–4 before he had completed his task, and his work was not published until 1791. His manuscripts are now in the Bodleian Library at Oxford.

It was at King's Cliffe that William Law, son of a grocer, Thomas Law, was born in 1686. He became a sizar of Emmanuel College, Cambridge, in 1705, and was made a fellow of his college in 1711. He was ordained and held several appointments, but most of his energies were thrown into literary work of a devotional character. In 1726 the first of his writings on *Christian*

Perfection appeared. An anonymous donor is said to have given him £1000 for this, and it has been suggested that it was with the money so obtained that he founded a school for fourteen girls at King's Cliffe. In the year 1728 he published the best known of his works—the *Serious Call*. Fifteen years afterwards, in company with two ladies of means, he settled at Thrapston with the purpose of carrying out literally the precepts of his great work; but in the following year the trio moved to Law's native place, King's Cliffe, where his ideal plan of life was adhered to strictly. Several other buildings were presented to the village by Law and his friends; these included a school for eighteen boys, almshouses, and a school building. His literary activity continued until his death in 1761.

Another devotional writer, whose works were at one time held in as high esteem as those of Law, was James Hervey, who was born at Hardingstone in 1713–14. His father was the incumbent of Collingtree. He was educated at the free Grammar School of Northampton and at Lincoln College, Oxford. At the University he met John Wesley, who prevailed upon him to learn Hebrew. He held various curacies and finally received his father's living. His writings, the best known of which is *Meditations among the Tombs*, were much admired in his day. He died in 1758.

William Paley is a name well known to all Cambridge men. His father was vicar of Helpston and minor canon at Peterborough. Born at the latter place in 1743, Paley entered Christ's College, Cambridge, in 1759, was Senior Wrangler in 1763, and in 1766 was elected a fellow of his college. He held several livings in Cumberland and Westmorland, became a prebendary of Carlisle in 1780, and two years later was made archdeacon. He published several books, the best known being *Horae Paulinae* which appeared in 1790, *The Evidences of Christianity* (still a text-book at Cambridge) which appeared in 1794, and *Natural Theology* which was published in 1802, three years before his death. In 1795 he had been made sub-dean of Lincoln and rector of Bishop Wearmouth.

Althorp Park is the home of the Spencers. George John Spencer, second Earl Spencer, was born in 1758 and was member of parliament for Northampton in 1780. Three years later his father died, and he succeeded to the earldom. He was a supporter of Pitt and became First Lord of the Admiralty in 1794. While he held this office the battles of St Vincent and Camperdown were won, and Nelson was sent out to the Mediterranean, where he gained the battle of the Nile. After being Home Secretary in 1806–7, Earl Spencer devoted himself to work in Northamptonshire and to the rehabilitation of the magnificent library at Althorp, at which place he died in 1834.

Apethorpe Hall is said to have passed into the hands of Sir Walter Mildmay (the founder of Emmanuel College, Cambridge) in the reign of Edward VI. His grand-daughter Mary inherited the property, and in 1624 married Sir Francis Fane, first Earl of Westmorland. Apethorpe remained the property of the Fanes until the beginning of this century. John Fane, eleventh Earl of Westmorland, better known perhaps as Lord

Burghersh, was born in 1784. He joined the army in
1803 and served in several campaigns, acting as aide-de-
camp to Wellington in the Peninsular War. He became

John Clare

a Privy Councillor in 1822, and was employed in many
diplomatic missions. He was promoted to the rank of
General in 1854. Not less distinguished as a musician
than as a soldier and diplomatist, he was the composer of

many musical works, and the moving spirit in the opening of the Royal Academy of Music in 1824. He died at Apethorpe in 1859.

James Thomas Brudenell, seventh Earl of Cardigan, was born in 1797. He entered the army and represented North Northamptonshire in Parliament in 1832. He went out to the Crimea in 1854 in command of a cavalry brigade, and led the famous charge at Balaclava. He died at his Northamptonshire seat, Deene Park, in 1868.

In 1793 John Clare, the "peasant-poet," was born in the parish of Helpston. His father was a poor labourer in receipt of parish relief, so that the son John gained what education he had with the greatest difficulty. A bookseller of Stamford first brought Clare's work before the public, and his first book of verse was successful, but the second was disappointing. Poverty pursued him all his life, and he died in the county lunatic asylum in 1864.

Another poet born in the county of humble parents was William Askham; he was born at Wellingborough in 1825, and began to learn shoemaking, his father's trade, when he was only ten years old. In the end he set up for himself, but he had found time meanwhile to educate himself, and he was made librarian of the Literary Institute at Wellingborough. He published, by subscription, four volumes of poems, and critics are agreed in praising the "fidelity of his nature poetry." He died at Wellingborough in 1894.

Born at Barnack Rectory in 1830, Henry Kingsley became, like his better known brother Charles, a notable writer of novels. He spent some years in Australia after leaving college, and his descriptions of Australian scenes still remain unexcelled. *Geoffrey Hamlyn* and *Ravenshoe*, his two best known books, hold a deservedly high place amongst English novels. For a short time he edited the

Thomas Littleton Powys, fourth Lord Lilford

Edinburgh Daily Review, but on the outbreak of the Franco-Prussian war he went out as war-correspondent to his paper, and was present at the battle of Sedan in 1870. He died six years later.

Bulwick has been the home of the Tryons since the

reign of James I. Vice-Admiral Sir George Tryon was born in 1832, and entered the navy at the age of sixteen. He had a distinguished career, and when he had attained the rank of Vice-Admiral was appointed to the Mediterranean command. In some evolutions carried out 22 June, 1893, his flagship the *Victoria* was rammed by the *Camperdown* and sank with the Admiral and 358 of her officers and crew.

A writer very popular in the second half of the nineteenth century, James Rice the novelist, was born at Northampton in 1843. He was called to the Bar but took up literary work, and from 1872 to 1882 collaborated with Sir Walter Besant. *Ready Money Mortiboy* was the first work they produced together, but the book for which they will always be best known is *The Golden Butterfly*, the humorous parts of which are generally ascribed to Rice's pen. Rice died in 1882, three years after the publication of his *History of the British Turf*.

Thomas Littleton Powys, fourth Baron Lilford, was born in 1833. Early devoted to the study of natural history, a traveller, and a first-rate linguist, he became the best field ornithologist in the kingdom and formed at Lilford his famous collection of living birds. He published much on ornithology. His great work— *Coloured Figures of the Birds of the British Islands*—began to appear in 1885, and was practically completed at the death of the author in 1896. In addition to various technical works he wrote a monograph on the *Birds of Northamptonshire and Neighbourhood* which was published in 1895.

In 1835 Oundle was the birthplace of a distinguished musician, Ebenezer Prout. He was a successful composer, but is best known as an authority on the theory of music. He succeeded Sir Robert Stewart as Professor of Music in the University of Dublin in 1894, and retained the post until his death in 1909.

22. THE CHIEF TOWNS AND VILLAGES OF NORTHAMPTONSHIRE.

(The figures in brackets after each name give the population of the town or parish in 1901, but in those cases where the figures for 1911 are available they are given and are distinguished by an asterisk. The figures at the end of the sections give the references to the text.)

Aldwinkle (370) is a small village on the Nene comprising two parishes (St Peter's and All Saints'). Both churches possess many points of interest, that of St Peter having a broach spire of particularly beautiful proportions. In the rectory of one (St Peter's) Thomas Fuller, the historian, was born in 1608; the rectory of the other (All Saints') was the birthplace of the poet John Dryden (1631). (pp. 30, 124, 126, 127, 129, 176, 177, 187.)

Apethorpe (170) is on the Willow Brook, about three miles north-west of Fotheringhay. Remains of a considerable Roman villa were discovered in the park here in 1858. In the village street is a well-preserved specimen of stocks and whipping-post. The Hall is of great interest; it was the home of Sir Walter Mildmay and served as a model for his foundation of Emmanuel College, Cambridge. (pp. 59, 85, 112, 148, 180, 182.)

Ashby St Ledgers (240) is situated on high ground four miles north of Daventry. It is interesting as the home of the Catesbys, and in the manor house close to the church may still be seen the room in which (according to local tradition) the Gunpowder Plot was arranged by Robert Catesby.   The church is Perpendicular and has its original roofs and rood screen.   (pp. 55, 127, 173.)

Aynho (445) is a small village at an elevation of more than 400 feet, six miles south-east of Banbury at the extreme south of

All Saints' Rectory, Aldwinkle
(Dryden was born in the room over the door)

the county.   The station of Aynho, on the G.W.R., lies at some distance from the village, in the valley of the Cherwell.   Mineral springs occur.   Whilst levelling an ancient road known as the Portway through the park of Aynho, some workmen found a skeleton, with the legs gathered up, enclosed between rough stone slabs.   (pp. 11, 107.)

NORTHAMPTONSHIRE

Barnack (614) is a village on the edge of the Fens three and a half miles south-east of Stamford, once famous for its quarries, which were situated in the broken ground now known as "Hills and Holes." The very interesting church (St John the Baptist) has a fine Saxon tower, showing "long-and-short" work, a large

Barnack Church

Early English south porch, and a Perpendicular Lady chapel. (pp. 76, 113, 118, 121, 123, 124, 161, 182.)

Barnwell (340) lies about two miles south of Oundle; there are many fine old houses and the remains of a castle. Originally there were two parishes here, but these were joined together in

1821. Of the two churches only the chancel of All Saints' remains; St Andrew's is Early English and Decorated, and contains many features of interest. One and a half miles to the west is Lilford Park, with a fine Tudor house, and celebrated for its wonderful aviaries and the collection of birds formed by the fourth Lord Lilford. (pp. 44, 136, 137.)

Blisworth (856) is on the main line of the L. & N.W.R. five miles south-west of Northampton; to which a branch line runs to Northampton. Ironstone is worked. (pp. 106, 157, 159.)

Brackley (*2633), on the river Ouse at the southern extremity of the county, has two stations (L. & N.W. and Gt Central). The church (St Peter) has an Early English tower and other points of interest. There is a secondary school (Magdalen College School). It is an old town, formerly of some importance, at one time having a considerable wool trade; it returned two members to parliament from the reign of Edward VI until 1832, in which year it was disfranchised. (pp. 27, 124.)

Braunston (854) is situated on the western edge of the hills in the west of Northamptonshire three and a half miles north-west of Daventry. The church (All Saints') was entirely rebuilt in 1849. Near it the Grand Junction Canal passes through the hills by a tunnel one and a half miles in length. (pp. 10, 151, 156.)

Braybrooke (269) stands on ground sloping towards the Welland valley, three miles south-east of Market Harborough. The church (All Saints') is Early English and Decorated; it has a fine spire and contains some interesting monuments and a British font. In the village is a Gothic fourteenth century bridge.

Brigstock (1000) is a village on Harper's Brook five miles north-west of Thrapston, taking its name from the ancient bridge which spans the stream. The Saxon tower of the church (St

Andrew) is of especial interest.   There is also a fine old market
cross in the centre of the village.   The kennels of the Woodland
Pytchley Hunt are here.   Brigstock was once a manor in the heart
of old Rockingham Forest and the country round is still thickly
wooded.   The manor house is of great interest and still retains
much of its fifteenth century character.   (pp. 32, 59, 67, 68, 113,
118, 120, 121.)

Brington, Great, (644) stands on a spur seven miles north-
west of Northampton.   The church (All Saints') is mainly Early
English, but the clerestory of the nave, the chancel, and the
north chapel are Perpendicular, and are generally attributed to
the architect of Henry VII's chapel at Westminster.   The church
contains many fine monuments to the Spencers.   (pp. 124, 127.)

Brington, Little, is a hamlet about a mile south of Great
Brington; its sole point of interest is a small house still standing
which is said to have been the home of the Washingtons of
Sulgrave when they moved to Brington at the beginning of the
seventeenth century.

Brixworth (1123) is situated on high ground (400 ft.)
seven miles north of Northampton overlooking the valley of the
tributary which joins the Nene at Northampton.   The church of
All Saints' contains much Roman material and was at one time
thought to be Roman work.   The hounds of the Pytchley Hunt
are kennelled here.   (pp. 106, 112, 113.)

Burton Latimer (2774) is situated two and a half miles
south-east of Kettering, overlooking the valley of the Ise; it
contains a large and interesting church possessing examples of all
the different styles of architecture from Norman to Perpendicular.
Iron ore mines are worked near.

Byfield (792), at a height of nearly 500 ft., is situated on
the old turnpike road midway between Daventry and Banbury.

Its church (Holy Cross) belongs to the second half of the four-teenth century. (p. 126.)

Canons Ashby (64) stands on high ground nine miles south of Daventry. A Priory was founded here before the time of Henry II, but the greater part of the buildings have now disappeared. In 1665 the Priory lands came into the possession of the Drydens, relatives of the poet. The manor house is of great interest; it was partly built by the first John Dryden about the middle of the sixteenth century. (pp. 87, 124, 126, 148.)

Castle Ashby (256), on the edge of the Nene valley, shares a station with Earl's Barton on the branch line between Peter-borough and Northampton from which latter it is distant seven miles eastward. Here is the seat of the Comptons, which is considered to be more finely situated than any other great house in the county. To the south lies Yardley Chase. (p. 144.)

Castor (639) is a village five miles west of Peterborough in the Nene valley where the latter opens on to the Fen plain. The kennels of the Fitzwilliam hounds are near here, in Milton Park. There was a considerable Roman town here (see Chapter 15), situated on Ermine Street at the point where King Street branched off, and the Romans had great pottery works in the town. Some of the kilns have been excavated, and many specimens of Castor ware have been found in different parts of the county. The church (St Kyneburgha) has a fine Norman tower. (pp. 84, 85, 106, 109, 111, 112, 121, 153, 192.)

Cold Ashby (258) probably derives its name in part from its exposed position; it is situated at a height of over 600 feet, near the source of Stowe Brook, 12 miles north-west of Northampton. The church (St Denis) is thought to be Norman in origin, and has a very old bell dated 1317. (pp. 21, 87.)

Colly Weston (361) is a small village overlooking the Welland valley, on the north-west edge of the ridge of high

Castor Church

ground which runs along the north side of the county. It is famous for its slate quarries, which have been worked since very early times. (pp. 33, 79, 80, 149, 154.)

Corby (1022), three miles south-east of Rockingham, situated on fairly high ground (over 300 feet), with a station on the M.R., produces large quantities of ironstone. (pp. 55, 81, 106, 126.)

Cotterstock (151), two miles north of Oundle, has a picturesque Hall which is very interesting on account of its associations with Dryden. The church (St Andrew) is mainly Early English; the choir and aisles are Decorated and the south porch Perpendicular. (pp. 30, 59, 112, 115, 124, 126.)

Cranford (two parishes—456) four and a half miles east-south-east of Kettering, lies at the head of a small valley which opens on to that of the Nene just above Thrapston; it has two churches (St Andrew and St John) now united under one rectory. There are extensive iron-works here.

Daventry (*3517) is a municipal borough and market town situated on high ground between the basins of the Nene and the Leam. North of the town are two reservoirs which supply water to the Grand Junction Canal, and immediately to the east of it rises Borough Hill (658 feet), on which some ancient earthworks can be traced (see Chapter 15). The town was of some importance in the old days of travelling by coach, as it stood at the junction of four important main roads, and it is said that at one time no fewer than 80 coaches passed through it daily. At that time it possessed a considerable manufacture of whips, but with the decline of coaching this industry died away, and the manufacture of boots and shoes took its place. The church (Holy Cross) was rebuilt in the second half of the eighteenth century. Charles I slept at the "Wheatsheaf" just before the battle of Naseby, and his army encamped on Borough Hill. (pp. 15, 17, 49, 73, 74, 89, 108, 109, 112, 151, 156, 157.)

Deene (123) is an extremely pretty village standing just off the main road from Kettering to Stamford some five miles east of Rockingham. The church (St Peter) was practically rebuilt in 1861, but the tower and spire and the main arcade of the nave, which are Early Decorated, were not touched. The hall and some other parts of the house which stands close to the church date from the reign of Edward VI. (pp. 129, 139, 144, 148, 182.)

Desborough (*4093) is situated five miles south-east of Market Harborough at a height of over 400 feet, overlooking the upper part of the Ise valley. The church (St Giles) is mainly Early English. Iron ore is obtained and the manufacture of boots and shoes is carried on. (pp. 106, 159.)

Duston (975) is situated on rising ground to the west of Northampton. Many interesting finds have been made here by antiquaries. (pp. 106, 112.)

Earl's Barton (2914) stands three miles south-west of Wellingborough upon a curving ridge of high ground projecting into the Nene valley. The church (All Saints) is celebrated for its massive Saxon tower, rudely built, with "long-and-short" work. There is a considerable amount of boot and shoe making. (pp. 36, 113, 117, 118, 121.)

Easton Maudit (121), situated about two miles north-east of Yardley Hastings, has an interesting church (St Peter and St Paul) with a fine spire; it contains some good monuments to the Yelvertons. (p. 126.)

Fawsley (59) stands on high ground three miles south of Daventry. There is a fine house here belonging to the Knightleys; it is of various dates, but the hall and kitchens may date back as far as the beginning of the sixteenth century. The hall has a fine oriel, above which is a small chamber in which some of the Martin Marprelate tracts are said to have been printed. The church (St Mary) contains some curious bench ends, glass dating from the sixteenth century, and many monuments to the Knightleys. (pp. 59, 140.)

Finedon (*3782) overlooks the valley of the Ise, three miles north of Wellingborough. The church (St Mary) is wholly Decorated and possesses a peculiar "strainer" arch at the east end of the nave. A great deal of ironstone is worked in the neighbourhood. (pp. 36, 75, 83, 126.)

Stocks and Whipping Post, Gretton

Fotheringhay (195) is a small village about two miles from Elton station, situated at the end of a "peninsula" which juts out into the valley of the Nene. The church (St Mary and All Saints) is a fine example of Perpendicular work and its beautiful lantern tower is a conspicuous landmark in the district. The castle has been described in Chapter 17. (pp. 22, 59, 89, 103, 127, 130, 133, 134, 135, 173, 175.)

13—2

Geddington (912) is a village in the valley of the Ise placed at the point where the stream turns south, about four miles north-east of Kettering. The church (St Mary Magdalene) possesses several points of interest. Standing within the limits of Rocking-ham Forest, the village was once a famous hunting resort of Henry II, and two great councils were held here (1177 and 1188). In the village, where three roads meet, stands a beautiful Eleanor cross, with three figures of the Queen. Part of the bridge over the river is of thirteenth century date. (pp. 87, 88, 160.)

Grafton Regis (138) stands on the road from Buckingham to Northampton four and a half miles south-east of Towcester. It was the home of Elizabeth Woodville, who was married privately to Edward IV in the manor house on May 1st, 1464. (p. 171.)

Gretton (777) is a small and picturesque village on the north-west border of the county, close to Rockingham and over-looking the valley of the Welland. The old stocks and whipping-post are still to be seen on the village green. To the south-east lie the ruins of Kirby Hall, a fine Elizabethan house, described in Chapter 18. (pp. 105, 106, 160, 195.)

Hargrave (240) lies on the eastern border of the county about two miles from Raunds on the Kimbolton road. During the restoration of the church in 1868 the interesting Hargrave Stone was found in some Early English masonry. The stone is roughly scratched for the game described in Shakespeare's *Midsummer-Night's Dream*, II. 2 as "Nine-men's morris." The game —a kind of draughts—is still played in some parts of the country.

Harringworth (284) is a village five miles north-east of Rockingham, in the valley of the Welland. Here the M.R. line from Kettering to Oakham crosses the valley by a long and lofty brick viaduct. There is a village cross, and several of the buildings possess points of architectural interest. (p. 160.)

Higham Ferrers Church and Archbishop Chichele's
School House

Helmdon (516) is situated on high ground near the source of the Tove midway between Towcester and Banbury. It was at one time celebrated for its stone quarries from which the building materials for many churches in the Midlands were obtained. The church (St Mary Magdalene) is a good example of late thirteenth century work.

Higham Ferrers (*2726), the birth-place of Archbishop Chichele, is a very old town situated on the south side of the

Irthlingborough Cross

Nene valley. It has been a municipal borough from very early times, and returned a member to Parliament until the year 1832. The church (St Mary) is of especial interest, belonging in the main to the Early English and Decorated periods. There are, in addition, an old school-house and bede-house (Perpendicular), and remains of Archbishop Chichele's College. The town consists of one long straight street, and has a considerable industry in boot

and shoe making.   There are two railway stations; that of the
M.R. (connecting it with Wellingborough) is in the town itself;
the other, on the L. & N.W.R., is at a little distance, in the Nene
valley, on the branch line from Northampton to Peterborough.
(pp. 72, 124, 154, 170, 175, 197, 199.)

Holdenby or **Holmby** (200) lies six and a half miles
north-west of Northampton.   Here once stood the Elizabethan

Thirteenth century Bridge over the Nene between
Irthlingborough and Higham Ferrers

manor house in which Sir Christopher Hatton was born.   Very
little of the original building was left when the house was restored
towards the end of the nineteenth century.   Charles I was brought
here and stayed in the house until he was removed from it by
Cornet Joyce.   (pp. 39, 89, 143, 175.)

Irchester (2301) is situated on the border of the county,
south-east of Wellingborough, in a side valley opening on to

that of the Nene. The name of the village shows its Roman origin (see Chapter 15). The church (St Katherine), belonging to the Decorated period, possesses a fine spire. Boot and shoe making is carried on, and ironstone is worked near. (pp. 85, 111, 159.)

Irthlingborough (*4630) is an industrial town on the north side of the Nene two miles north of Higham Ferrers, with important manufactures of boots and shoes, and large ironstone quarries. The church (St Peter) is of great interest, its most noteworthy feature being a semi-detached bell-tower. There are the remains of a fine market cross. The road to Higham Ferrers crosses the Nene by a fine thirteenth century bridge. (pp. 72, 83.)

Islip (600) is a picturesque village on the opposite side of the Nene to Thrapston, lying just off the road from Huntingdon to Kettering. The church (St Nicholas) is a beautiful example of Perpendicular work, with a fine spire. Horses' collars are manufactured, the outer covering being made of rushes obtained from the Nene valley. There are large ironworks near. (pp. 34, 75, 82.)

Kettering (*29,976) is an old and important town placed on high ground overlooking the valley of the Ise. The church (SS. Peter and Paul) is in the main Perpendicular. It has a magnificent tower and spire. The population in 1841 was 4867; in 1871 it had risen to 7184, in 1901 to 28,653, and in 1911 to 29,976. This increase has been due to the improved means of communication afforded by the construction of the M.R., and to the rise of an important industry in boots and shoes. The town has other manufactures, such as those of agricultural implements, clothing, and brushes, and a great deal of iron is obtained in the vicinity. There are many fine buildings and a grammar school. (pp. 21, 35, 59, 72, 75, 78, 80, 127, 129, 154, 159, 160, 167.)

King's Cliffe (983) is a village in the valley of the Willow Brook. Here William Law, author of *The Serious Call*, was born and lived. There is an old industry, still carried on, of wood carving and turning. (pp. 105, 109, 161, 178, 179.)

King's Sutton (1037) lies in the valley of the Cherwell, 4½ miles south-east of Banbury. It has a fine church (St Peter) which is mainly Early Perpendicular; the best features are the tower, spire, and west front. The manor house is picturesque, and is said to have once afforded a refuge to Charles I. There are some tumuli in the parish, and Roman coins have been found in great quantities. (pp. 105, 127.)

Kislingbury (649) is a village three miles west of North-hampton, on the Nene. The church (St Luke) is in the Decorated style. (p. 35.)

Long Buckby (2147) is situated five miles north-east of Daventry on high ground (over 400 ft.) to the east of the spot where the Grand Junction Canal crosses the old line of Watling Street. A great many hands are employed in the manufacture of boots and shoes.

Longthorpe (282) is a hamlet two miles west of Peter-borough. Here there is a house which dates from the latter part of the thirteenth century, interesting as showing a small keep or peel-tower attached to it. The house once formed the summer palace of the Abbot of Peterborough. (p. 139.)

Lowick (335) is on Harper's Brook, a short distance above its confluence with the Nene. The church (St Peter) is Perpen-dicular, with a fine lantern tower, and possesses many points of interest; the glass and the monuments deserve mention. In a field to the south of the village is "the Lowick Oak," one of the largest trees of its species in the kingdom. South-west of the village is situated Drayton House. (pp. 31, 113.)

Middleton Cheney (1057) is a village on high ground (over 400 ft.) two and a half miles north-east of Banbury. The church (All Saints) is mainly Decorated, with a Perpendicular tower. In 1643 a fight took place in a neighbouring field between the Royalists and Parliamentarians.

Moreton Pinkney (378) is situated on high ground (450 ft.) eight miles west of Towcester near one of the sources of the Cherwell. To the north lies Canons Ashby, an old house of great interest. (p. 124.)

Naseby (476) stands higher than any other village of the county, with the exception of Cold Ashby. To the north of the village lies Naseby Field, where Charles I was defeated 14th June, 1645. (pp. 17, 21, 26, 49, 55, 87, 89, 175.)

Nassington (505) is a small village in the valley of the Nene near the Huntingdon border. The church (St Mary and All Saints) is a very interesting one, showing work of several periods. The old Prebendal house is now used as a farm dwelling. (pp. 59, 115, 126.)

Northampton (*90,076), a municipal, county, and parliamentary borough, the county town, lies on a slope on the north bank of the Nene, 69 miles from London by road. It returns two members of Parliament. It is on the London and North Western Railway, and on a branch line of the Midland.

The history of Northampton has already been given (p. 90). The great fire of 1675, which burnt over 600 houses, has made it mainly a modern town, and not many of the old houses exist. It is a busy place, the depôt of the Northamptonshire Regiment, with an enormous trade in boots and shoes, and large leather and tanning factories besides foundries and paper-mills. It is composed of four parishes: All Saints', St Giles's, St Peter's, and St Sepulchre's. The church of All Saints perished in the fire and the present building was erected not long afterwards by Wren. St Peter's, probably built about 1160, on the site of an early Saxon church, shows very rich Norman work of remarkable interest, but a scarcely less interesting feature of the town, perhaps, is the church of the Holy Sepulchre, one of the four "Round Churches" existing in England, the others being at Little Maplestead, and Cambridge, and the Temple Church in London. This has a chancel with lateral chapels and a circular nave with eight Norman shafts bearing pointed arches. At the west end is a Perpendicular tower with a spire, supported by buttresses of colossal size.

Oundle School Hall

The Town Hall is a fine but modern building; but the County Hall was built after the 1675 fire and has some good pargetting. The Museum contains a quantity of Romano-British remains found at Irchester and Towcester and also some of the British objects excavated at Hunsbury Camp.

About two miles from the town, in the parish of Hardingstone, stands one of the three remaining Eleanor Crosses. (pp. 2, 3, 19, 21, 55, 56, 59, 70, 72, 73, 75, 78, 80, 83, 85, 87, 90, 92—96,

105, 107, 108, 112, 121, 124, 126, 127, 132, 151, 154, 155, 157, 158, 159, 160, 164, 165, 167, 169, 174, 176, 179, 180, 184.)

Northborough (198) lies in the Fen district two miles south-east of Market Deeping. The church (St Andrew) is mainly fourteenth century work, and has a beautiful chapel, and the manor house is of very considerable interest in spite of modern alterations. (pp. 140, 141.)

Great Oakley (200) is situated near the source of Harper's Brook, five miles north of Kettering. To the west of the village is the site of Pipewell Abbey, where Richard I held the first great council of his reign in 1189. The abbey was demolished after the dissolution of the monasteries.

Oundle (*2749) is a small market town overlooking the Nene valley. The church (St Peter) has a magnificent spire which serves as a landmark for miles round. The town has preserved its old-fashioned character; the houses, built of stone and roofed with Colly Weston slates, present many points of interest. In the year 709 Bishop Wilfrid of Hexham died here while making a round of visits to the monasteries which he had founded. There is a large public school with many fine buildings; there is also a grammar school on the same foundation. Brewing is the only industry. (pp. 7, 22, 44, 48, 59, 103, 106, 115, 124, 126, 127, 128, 129, 147, 174, 185, 203.)

Peterborough (*33,578), a municipal and parliamentary borough, head of the Soke of Peterborough, declared a separate "Administrative County," distinct from the rest of Northants, which comprises about thirty parishes besides Peterborough itself. The city stands on the north bank of the Nene, 76 miles from London by rail and 81 by road, partly in Huntingdonshire, and at the junction of four important railway systems—the Great Northern, Great Eastern, Midland, and London and North Western. It returns one member only.

Standing in the centre of a great agricultural district and a plexus of railway systems the industries and trade are necessarily such as appertain to them, and there are large markets for corn and stock, and fairs for cattle and horses, horse and dog shows, etc., besides factories of agricultural implements ; while there are also large engineering works, and considerable trade in coal and timber, bricks, and boots and shoes.   The Guildhall was built in 1671 ; the lower part of the building serving as a butter market, the upper, carried on shafts bearing round-headed arches, being the chamber used for municipal business.   In the Museum is a collection of peculiar interest—ships and other objects, beautifully modelled, made by the French prisoners at Norman Cross at the beginning of the last century from their beef bones, straw, etc.

The great feature of Peterborough is the cathedral, built entirely of Barnack stone and showing the most magnificent west front of all our English cathedrals.

In 655 a monastery was founded at Medeshamstede (afterwards Peterborough) by Saxulf, and its church was dedicated to St Peter, St Paul, and St Andrew.   The Danes however plundered and burnt this church in 870, and for nearly a century it remained in ruins.   Ethelwold, Bishop of Winchester, rebuilt it in 966, but this Saxon church, which plays a considerable part in the story of Hereward, was destroyed by fire in 1116.   Foundations of part of it have been discovered under the south transept of the cathedral.   The present church was begun by Abbot John of Sais, the new choir being finished in 1140 ; the transepts and nave are magnificent examples of Norman work ; the massive piers, round arches, and great length of the nave are specially worthy of notice.   In the Early English period the monks added the West Front with its three great arches and flanking towers ; this work has been much criticised, but it is considered by many to be the finest feature of the cathedral.   The Lady Chapel was built in the Geometrical period, but it was pulled down in the

Rockingham Village

seventeenth century—apparently for the sake of its materials. In the Decorated period the Norman tower was taken down for constructional reasons, and a new and lighter tower was built in its place. In the next period—the Perpendicular—the Galilee porch was inserted between two piers of the West Front. In 1541 the church became the cathedral of a new diocese, and the last Abbot, John Chambers, became the first Bishop of Peterborough. (pp. 9, 11, 13, 19, 22, 43, 59, 64, 70, 76, 77, 85, 90, 97—103, 106, 109, 123, 124, 129, 139, 145, 154, 155, 158, 160, 161, 164, 167, 168, 169, 179.)

Polebrook (345) lies in a valley running into the Nene, two miles south-east of Oundle. Its church (All Saints) is of great interest throughout; the tower and spire (Early English) are most perfectly proportioned; the nave is late Norman. (pp. 124, 129.)

Pytchley (547) is a small village three miles south of Kettering on high ground overlooking the valley of the Ise. Formerly the headquarters of the Pytchley hunt were here.

Raunds (*3874) is a small town four miles south of Thrapston situated in a little side valley opening on to that of the Nene. The church (St Peter) is of great interest, its Early English tower and spire being particularly fine; the rest of the church is mostly Decorated. The manufacture of boots and shoes is the principal industry. (pp. 72, 124, 159.)

Rockingham (157), clinging to the slope of the ridge which descends abruptly to the Welland valley, is one of the most picturesque villages in the county. Dominating the village is Rockingham Castle (see Chapter 17), of great historical interest. The village gave its name to what was one of the largest of English forests, extending from Northampton to Stamford and from the Welland to the Nene. (pp. 17, 38, 69, 87, 132, 135, 175, 206.)

Interior of Rushden Church
(*Showing "Strainer" arch*)

Rothwell (*4416) stands four miles north-west of Kettering on high ground overlooking the valley of the Slade, a small tributary of the Ise. The church is said to be the longest in the county; it contains work of many periods. In a crypt beneath there is a great collection of human bones. The town has manufactures of boots and shoes, and of clothing. The market-house is a striking feature of the town, and is one of the fantastic buildings of Sir Thomas Tresham. (pp. 106, 124.)

Sulgrave Manor, the Home of the Washingtons

Rushden (*13,354) lies close to Higham Ferrers in a small valley opening on to that of the Nene near Wellingborough. The church (St Mary) has a very graceful tower and spire (late Decorated) and other interesting features, amongst which is a "strainer" arch similar to one at Finedon. The increase in population from 2122 in 1871 to over 13,000 in 1911 is due to the development of the boot and shoe industry in the town. The M.R. branch from Higham Ferrers to Wellingborough passes through it. (pp. 30, 60, 72, 154, 208.)

Rushton (503) is a small village in the valley of the Ise three and a half miles north-north-west of Kettering. The Hall, once the home of the Treshams, and the triangular Lodge, built by Sir Thomas Tresham, are buildings of great interest. (pp. 21, 144, 160, 170, 171, 172.)

Stamford Baron (243) is part of the town of Stamford lying south of the Welland. The church (St Martin's) is entirely Perpendicular. On the hill above stands Burghley House, the home of the Cecils. (pp. 15, 17, 26, 80, 127, 153, 154, 160, 161, 174, 182.)

Stanford (35) is on the Leicestershire border about seven miles from Rugby. It has an interesting church (St Nicholas), chiefly remarkable for its ancient stained glass windows and beautiful monuments.

Stowe Nine Churches (242) is a small village on high ground near Watling Street. The tower of the church (St Michael and St James) is sometimes claimed as an example of Saxon work. The origin of the peculiar name of the village is not known. (p. 159.)

Sulgrave (395) stands on high ground (over 450 ft.) in the south-west of the county about ten miles west-south-west of Towcester. The chief point of interest is the Washington Manor House, the home of the ancestors of George Washington. (pp. 27, 209.)

Thorpe (176) is a small village in the valley of the Nene about three miles north-east of Thrapston. It contains some picturesque old houses and the remains of a castle built by the family of Waterville. (pp. 59, 137, 176.)

Thrapston (1747) is a small market town situated in the valley of the Nene on the road from Huntingdon to Leicester. It is an important agricultural centre and has two railway stations, one on the L. & N.W. branch from Northampton to Peterborough, the other on the M.R. branch from Huntingdon to Kettering. (pp. 21, 36, 70, 72, 124, 126, 127, 154, 159, 179.)

Towcester (2371) is a picturesque old town on the river Tove, built on the site of a Roman station (Lactodorum) on Watling Street. The church (St Lawrence) has some interesting features. Pillow-lace making and the manufacture of boots and shoes are the principal industries. The town was fortified by Prince Rupert in 1693, and was the only Royalist stronghold in Northamptonshire. (pp. 27, 84, 85, 105, 109, 111, 124, 151, 159, 173.)

Walgrave (682) is a village on rising ground seven miles north-west of Wellingborough. The church (St Peter) is Decorated and has a fine spire. Boot-making is carried on.

Wansford (67) is a very small village on the Nene on the borders of Huntingdonshire eight miles west of Peterborough. The church (St Mary) is of Saxon origin, and has a remarkable Norman font. The station, about a mile distant, is on the Northampton to Peterborough branch of the L. & N. W. R.; there are also branches to Stamford by Barnack, and to the valley of the Welland by King's Cliffe. The place had some importance in the old coaching days on account of its situation on the Great North Road at the point where it crossed the Nene. (pp. 18, 22, 151, 161.)

Warkton (233) is a small village on the Ise, two miles north-east of Kettering. The church (St Edmund) has a Perpendicular tower, but the chancel was rebuilt in the eighteenth century; it contains the monuments of the Montagus.

Warmington (547) is situated on rising ground just off the main road from Oundle to Peterborough two and a half miles north-east of the former. The church (St Mary) affords most beautiful examples of Early English work in all stages. The original nave roof, vaulted in wood, is one of its many interesting features. (pp. 30, 124, 125, 129.)

Weedon Beck (1868) is situated in the upper valley of the Nene at the point where Watling Street crosses the river four miles south-east of Daventry; the L. & N. W. main line has a station here, and the Grand Junction Canal passes through the village. A military depôt was established here in 1803, the spot being chosen chiefly on account of its central position. (pp. 21, 49, 56, 151, 159.)

Weldon (**Great** 280, **Little** 472) lies at the head of the valley of the Willow Brook eight miles north-east of Kettering. The freestone quarries have been famous for centuries. The church (St Mary) of Great Weldon (or Weldon-in-the-Woods) has a tower surmounted by a lantern, which served in old times as a lighthouse to guide travellers through that part of the forest of Rockingham. (pp. 31, 105, 106, 112, 160, 175.)

Wellingborough (*19,758) is situated at the junction of the Ise and the Nene. The church (St Luke) is of some interest, mainly Decorated and Perpendicular. In 1871 the number of inhabitants was 9385, so that in 40 years the population has doubled. This has been mainly due to the growth of the boot and shoe industry. The town is served by two railways; the Midland main line crosses the valley of the Nene here and has its station in the town; the L. & N. W. branch from Northampton to Peterborough has a station at a little distance from the town. There is an important Grammar School. (pp. 21, 36, 59, 72, 75, 126, 127, 159, 182.)

Yardley Hastings (958) lies on the main road from Bedford to Northampton three miles south of Castle Ashby. The church (St Andrew) belongs chiefly to the Decorated period. The remains of an old manor house stand near. Yardley Chase lies to the south-west. (p. 38.)

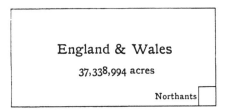

Fig. 1.   The area of Northamptonshire (638,612 acres)
compared with that of England and Wales

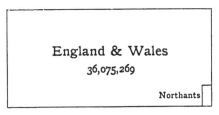

Fig. 2.   The population of Northamptonshire (348,552) compared with that of England and Wales (1911)

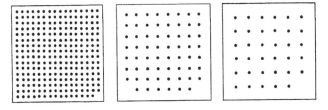

Lancashire 2550   England and Wales 618   Northants 349

Fig. 3.   Comparative Density of Population to
square mile (1911)

(Each dot represents ten persons)

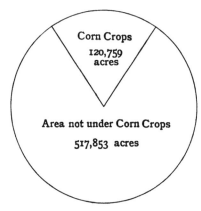

Fig. 4. Proportionate area under Corn Crops in
Northamptonshire (1909)

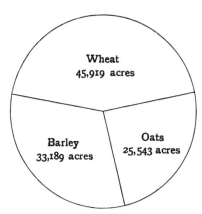

Fig. 5. Proportionate area of Chief Cereals in
Northamptonshire (1909)

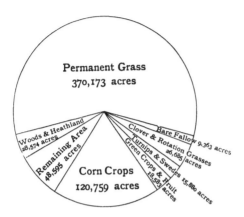

Fig. 6. Proportionate area of Permanent Pasture, Corn
Crops, etc., in Northamptonshire (1909)

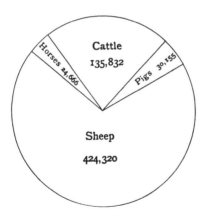

Fig. 7. Proportionate numbers of Live Stock in
Northamptonshire (1909)

www.ingramcontent.com/pod-product-compliance
Ingram Content Group UK Ltd.
Pitfield, Milton Keynes, MK11 3LW, UK
UKHW042143280225
455719UK00001B/53

9 781107 630987